A-Level Pass Book

CAMBRIDGE MODULAR EXAMINATIONS

CHEMISTRY

E. N. Ramsden
B.Sc., Ph.D., D.Phil.
Formerly of Wolfreton School, Hull

Stanley Thornes (Publishers) Ltd.

© E.N. Ramsden 1996

The right of E.N. Ramsden to be identified as author of this work has been asserted by her in accordance with the Copyright, Designs and Patents Act 1988.

All rights reserved. No part of this publication may be reproduced or transmitted in any form or by any means, electronic or mechanical, including photocopy, recording, or any information storage and retrieval system, without permission in writing from the publisher or under licence from the Copyright Licensing Agency Limited. Further details of such licences (for reprographic reproduction) may be obtained from the Copyright Licensing Agency Limited, of 90 Tottenham Court Road, London W1P 9HE.

First published 1996 by
Stanley Thornes (Publishers) Ltd
Ellenborough House
Wellington Street
CHELTENHAM
GL50 1YD

A catalogue record for this book is available from the British Library.

ISBN 0 7487 2436 2

Typeset by Techset Composition Ltd, Salisbury
Printed and bound in Great Britain

Contents

Preface v | Notes on terms used in examination papers vii

SYLLABUS MODULE 4820: FOUNDATION (Compulsory module)

Concept Maps: Foundation	1	Additional Topic Foundation 2: Practice on type of bonding	11
Module 4820: Foundation	8	Additional Topic Foundation 3: Period 3	11
Additional Topic Foundation 1: Practice on stoichiometric relationships	10	Examination questions on Foundation Module	17

SYLLABUS MODULE 4821: CHAINS AND RINGS (Compulsory module)

Concept Maps: Chains and Rings	20	Additional Topic Chains and Rings 2: The hydrolysis of proteins	28
Module 4821: Chains and Rings	26	Additional Topic Chains and Rings 3: Recycling materials	29
Additional Topic Chains and Rings 1: Commercial uses of esters	28	Examination questions on Chains and Rings	31

SYLLABUS MODULE 4822: TRENDS AND PATTERNS

Concept Maps: Trends and Patterns	34	Additional Topic Trends and Patterns 2: Improved batteries	41
Module 4822: Trends and Patterns	39	Additional Topic Trends and Patterns 3: Ceramics	43
Additional Topic Trends and Patterns 1: Determination of the Avogadro constant	40	Examination questions on Trends and Patterns	46

SYLLABUS MODULE 4823: MATERIALS

Concept Maps: Materials	48	Module 4823: Materials	51

SYLLABUS MODULE 4824: ENVIRONMENTAL CHEMISTRY

Concept Maps: Environmental Chemistry	53	Module 4824: Environmental Chemistry	58

SYLLABUS MODULE 4825: METHODS OF ANALYSIS AND DETECTION

Concept Maps: Methods of Analysis and Detection	60	Module 4825: Methods of Analysis and Detection	67

SYLLABUS MODULE 4826: HOW FAR, HOW FAST? (Compulsory module)

Concept Maps: How Far, How Fast?	69	concentration on electrode potential	77
Module 4826: How Far, How Fast?	74	Additional Topic How Far, How Fast? 4: Additional practice on reaction mechanisms	78
Additional Topic How Far, How Fast? 1: The effect of ionic charge and ionic radius on lattice energy	75	Additional Topic How Far, How Fast? 5: Enzymes	80
Additional Topic How Far, How Fast? 2: Method of measuring the standard electrode potential of ions of the same element in differerent oxidation states	76	Additional Topic How Far, How Fast? 6: The biological importance of buffer solutions	83
Additional Topic How Far, How Fast? 3: The effect of		Examination questions on How Far, How Fast?	84

SYLLABUS MODULE 4827: BIOCHEMISTRY

Concept Maps: Biochemistry	88
Module 4827: Biochemistry	94
Additional Topic Biochemistry 1: Hexokinases and Glucokinase	95
Additional Topic Biochemistry 2: Iron in haemoglobin and cytochrome	95

SYLLABUS COMPLEMENTARY MODULE 4843: FOOD TECHNOLOGY

Concept Maps: Food Technology	96
Complementary Module 4843: Food Technology	102
Additional Topic Food Technology 1: Quality of beef	103
Additional Topic Food Technology 2: Crop production	105
Additional Topic Food Technology 3: Fruits and vegetables	107
Additional Topic Food Technology 4: Ripening fruit	107
Additional Topic Food Technology 5: Strawberries spoiled by *Botrytis cinerea*	108
Additional Topic Food Technology 6: Potato crisps	108

Answers to Checkpoints on Additional Topics	109
Answers to examination questions	112
Index of additional topics	119

PREFACE

THE CAMBRIDGE MODULAR SYLLABUSES

This book is a guide to my text, *A-level Chemistry*, Third edition, E.N. Ramsden (Stanley Thornes Publishers). It indicates where in the text students will find each topic mentioned in the University of Cambridge Local Examinations Syndicate Modular Science syllabuses: A-level Chemistry 9525 and AS-level Chemistry 8525.

For some of the new modules the coverage in *A-level Chemistry* is incomplete. The reason is that when I wrote the text to include every topic in every Examining Board's syllabus the book turned out dauntingly long, and I deleted some topics which were not wanted by all Boards or which were examined only in special papers. In this guide I have included those items in the UCLES syllabuses which need extra coverage. I have added checkpoints, with answers, on the new material.

The guide will help students to prepare for their modular examinations. Students can focus on the parts of the syllabus which they wish to revise for a particular module.

TOPICS FOR CAMBRIDGE SYLLABUSES

For each module in Chemistry 9525 the guide gives a full list of references to *A-level Chemistry* or to my module books, *Biochemistry and Food Science*, *Materials Science*, *Chemistry of the Environment* and *Detection and Analysis*, and in some cases to Additional Topics within this guide.

CONCEPT MAPS FOR REVISION

Each syllabus topic has a concept map which illustrates the structure of the topic. The map also gives references to the sections in *A-level Chemistry* or a module book or an Additional Topic in this guide to indicate where each syllabus item can be found.
I hope that students will find these presentations helpful in planning their work and especially helpful for last-minute revision when a text book is too long to assimilate.

QUESTIONS FROM EXAMINATION PAPERS WITH ANSWERS AND COMMENTS FROM AN EXAMINER

I have included questions from Cambridge Modular examination papers. Some have outline answers and some have typical students' answers with an examiner's comments on how the students can improve their answers.

ACKNOWLEDGEMENTS

I thank Dr Rob Ritchie for his careful reading of the first draft and for the valuable comments and suggestions which he contributed.

I thank the University of Cambridge Local Examinations Syndicate for permission to reproduce questions from examination papers.

I thank the publishing team at Stanley Thornes for their expertise in preparing the material for publication.

I thank my family for their encouragement.

E.N. Ramsden,
Oxford, 1996

NOTES ON TERMS USED IN EXAMINATION PAPERS

You should note exactly what the question is asking for; if it asks you to 'explain' you are expected to give a longer reply than if you are asked to 'state'. The following list may help you.

1. **State** asks for a concise answer with little discussion or argument.

2. **Define** asks for only a formal statement.

3. **List** asks for a number of points, which may be only one word, with no amplification.

4. **Outline** asks for a brief account of essential points.

5. **Describe** asks you to state in words, with diagrams if these are helpful, the main points in a topic. If the question is about experiments, you should include visual observations.

6. **Explain** implies that you should give reasons and refer to theory.

7. **Discuss** asks for a critical account of a topic.

8. **Suggest** is used when there is more than one correct solution and also when you are expected to apply your knowledge to a novel situation.

9. **Predict** or **deduce** implies that you are not expected to recall the answer but to arrive at it by making connections between items of information given in the question.

10. **Comment** is an open-ended term, asking you to recall or infer relevant material. The mark allocation will guide you on the length of answer to give.

11. **What do you understand by?** implies that you should give a definition and also some comment on the significance of the term or terms. The mark allocation will guide you on how much detail to give.

12. **Sketch** means that a freehand drawing is acceptable. However, you should take care to show whether a graph passes through the origin or has an intercept and get the proportions approximately right.

13. **Calculate** and **find** and **determine** are used when a numerical answer is required.

14. **Units** must be quoted with physical quantities.

15. **Significant figures** should be quoted correctly: if your data are given to three significant figures, e.g. 25.0 cm^3, quote your answer to the same degree of accuracy, e.g. $1.25 \times 10^{-3} \text{ mol dm}^{-3}$, neither $1.2 \times 10^{-3} \text{ mol dm}^{-3}$ nor $1.258 \times 10^{-3} \text{ mol dm}^{-3}$.

4820: Foundation

1. ATOMS, MOLECULES AND STOICHIOMETRY

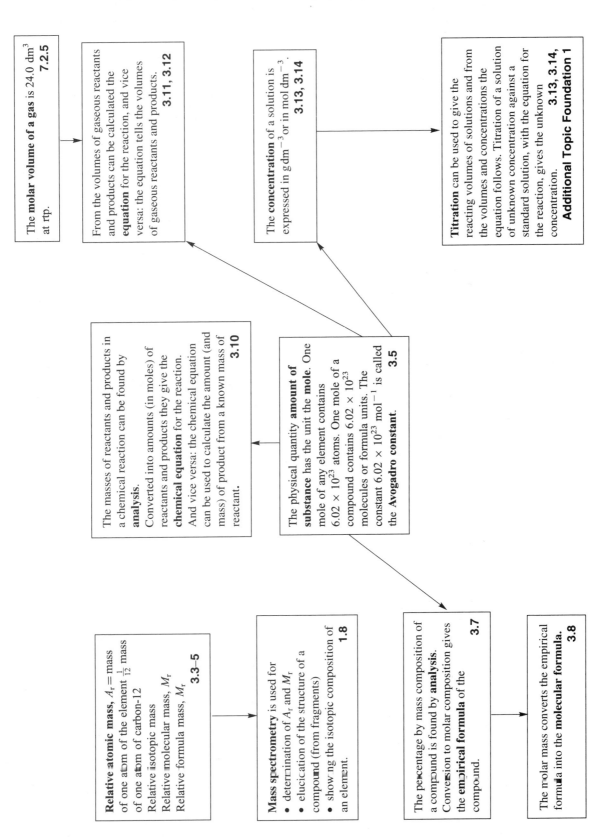

2. ATOMIC STRUCTURE

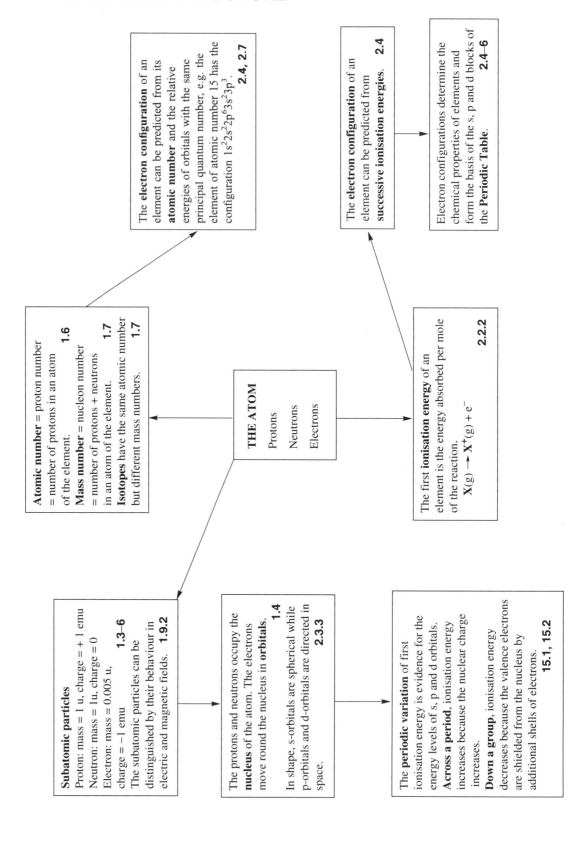

3. CHEMICAL BONDING

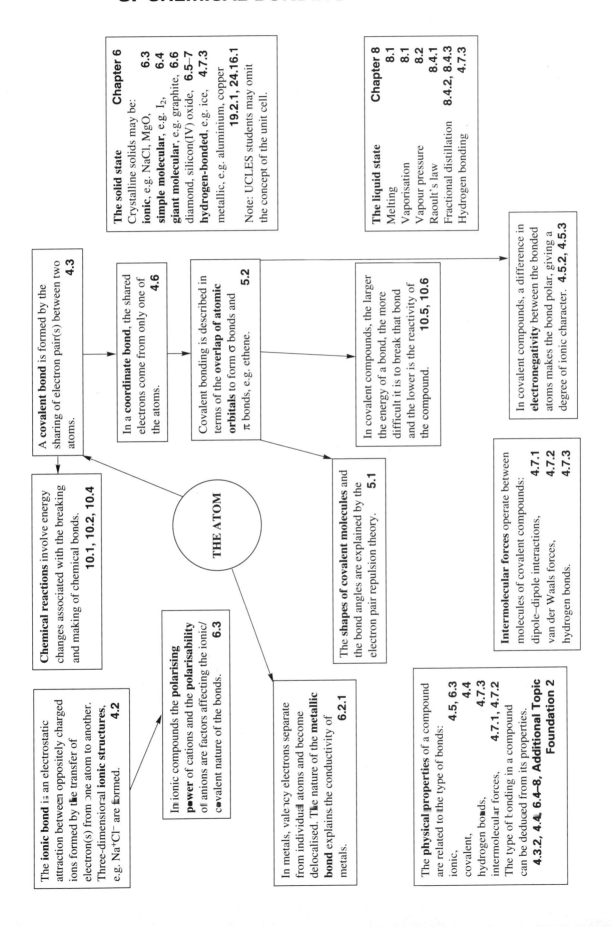

4. CHEMICAL ENERGETICS

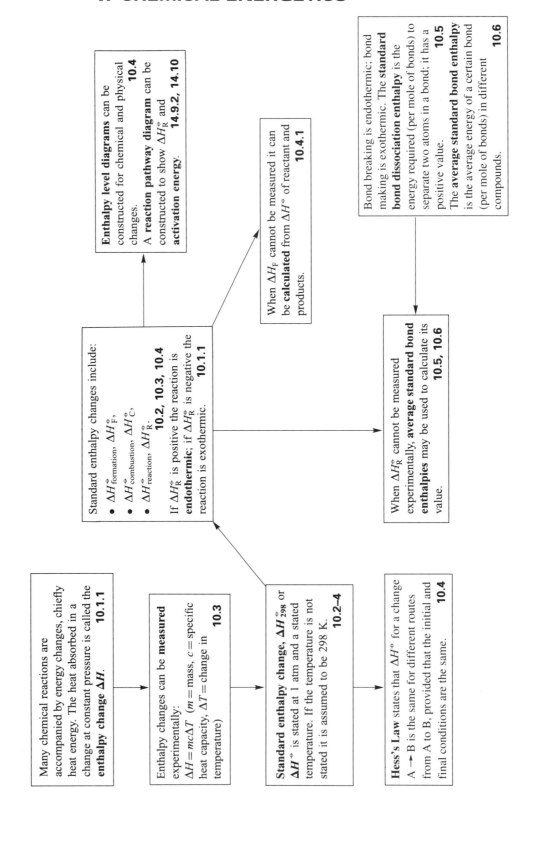

FOUNDATION

5. THE PERIODIC TABLE

PERIODIC TABLE

In the Periodic Table, elements are listed in order of **atomic number**. They are arranged in vertical **groups** and horizontal **periods**. **18.1**

Properties which vary in a periodic manner include
- atomic radius and ionic radius,
- first ionisation energy,
- melting temperature,
- electrical conductivity.
15.1 and Additional Topic Foundation 3

Down the groups:
Atomic radius and ionisation energy vary gradually. Bonding and valency change.
The first member of the group has atypical properties. **18.1**

The **Group 1** elements Li, Na, K, react with water in a vigorous, exothermic reaction to form the metal hydroxide and hydrogen. The vigour of the reaction increases down the group as the ionisation energy decreases. **18.1**

The **third period** is elements:
Na Mg Al Si P S Cl Ar
Chemical properties of the elements that change across the period include
- the type of bonding, which changes from ionic in Groups 1 and 2, with the element as cation, to ionic or covalent in Group 3 to covalent in Groups 4 and 5 to ionic in Groups 6 and 7, with the element as anion,
- the oxidation number of the element,
- the reaction with water,
- the reaction with oxygen,
- the reaction with chlorine.
15.2, 15.3, Additional Topic Foundation 3

Oxides and chlorides of Period 3
- The reactions of the oxides with water range from the strongly basic oxides of Groups 1 and 2 to the strongly acidic oxides of Groups 6 and 7. **15.3.1**
- The hydroxides of Groups 1 and 2 are strongly basic. **15.3.1**
- The chlorides of Groups 1 and 2 are ionic and dissolve in water to form neutral solutions. The chlorides of the non-metallic elements, except carbon, are hydrolysed to give acidic solutions. **15.3.2**

TRANSITION ELEMENTS

Transition elements or **d-block elements** have one or more stable ions with incompletely filled d orbitals. The **electron configurations** of the transition metals from Ti to Cu are Ti(Ar)$3d^24s^2$ to Cu(Ar)$3d^{10}4s^1$. **24.1**

Oxidising agents include MnO_4^-, MnO_2, $Cr_2O_7^{2-}$, Fe^{3+}, Cu^{2+}. **Reducing agents** include Fe^{2+}, Cu^+. Redox reactions used in **titration**.
3.15.1, 24.10.1, 24.10.2

Transition metals have variable **oxidation states**, e.g. Fe (+2, +3), Cr (+6, +3), Mn (+7, +6, +4, +2) and can change oxidation states on oxidation or reduction. **24.6**

They are widely used as **catalysts**, e.g. iron in the Haber Process, platinum in the Contact Process and platinum and rhodium in catalytic converters. **14.1.5, 24.7**

See **oxidation state** or **oxidation number**. **3.15–17**

Transition metals form **complex ions** by coordinate bonding to ligands, e.g. H_2O, NH_3, Cl^-, $C_2O_4^{2-}$, $H_2NCH_2CH_2NH_2$ and edta, all of which are **Lewis bases**. **24.13**

In **charge**, complexes may be cationic, e.g. $[Cu(NH_3)_4(H_2O)_2]^{2+}$ anionic, e.g. $[Zn(OH)_4(H_2O)_2]^{2-}$ uncharged, e.g. $Ni(CO)_4$. **24.13**
In **shape**, complexes may be square planar, e.g. $Ni(CO)_4$ octahedral, e.g. $[Cr(NH_3)_6]^{3+}$. **12.7.17, 24.13.5–9**
The **colour** of a complex depends on the oxidation state of the metal and on the identity of the ligand. **24.13.2, 24.13.9**
The **naming** of complex ions. **24.13.1**

Across a period, atomic radii, ionic radii and first ionisation energies vary little across a set of transition metals.
Compared with s- and p-block metals, transition metals are less reactive, harder, with higher melting temperatures. **24.1–3, 24.2.2**

6. CARBON COMPOUNDS: AN INTRODUCTION TO ORGANIC CHEMISTRY

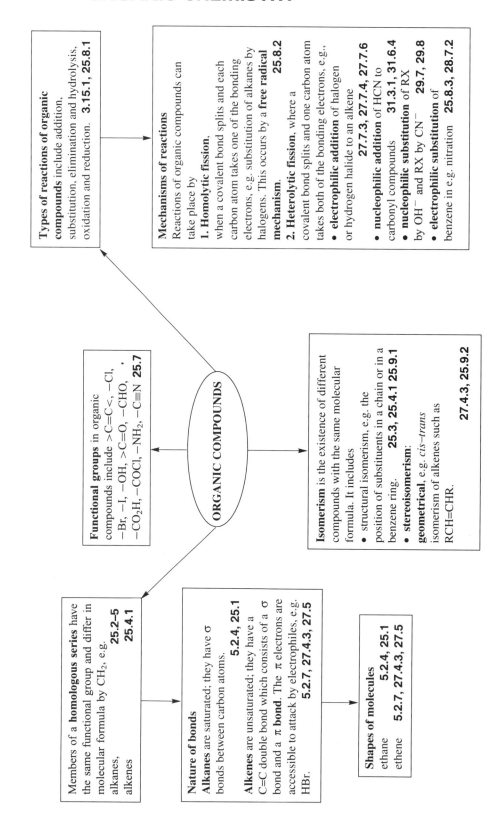

Types of reactions of organic compounds include addition, substitution, elimination and hydrolysis, oxidation and reduction. **3.15.1, 25.8.1**

Mechanisms of reactions
Reactions of organic compounds can take place by
1. **Homolytic fission**, when a covalent bond splits and each carbon atom takes one of the bonding electrons, e.g. substitution of alkanes by halogens. This occurs by a **free radical mechanism**. **25.8.2**
2. **Heterolytic fission**, where a covalent bond splits and one carbon atom takes both of the bonding electrons, e.g.,
- **electrophilic addition** of halogen or hydrogen halide to an alkene **27.7.3, 27.7.4, 27.7.6**
- **nucleophilic addition** of HCN to carbonyl compounds **31.3.1, 31.6.4**
- **nucleophilic substitution** of RX by OH^- and RX by CN^- **29.7, 29.8**
- **electrophilic substitution** of benzene in e.g. nitration **25.8.3, 28.7.2**

Functional groups in organic compounds include $>C=C<$, $-Cl$, $-Br$, $-I$, $-OH$, $>C=O$, $-CHO$, $-CO_2H$, $-COCl$, $-NH_2$, $-C\equiv N$ **25.7**

ORGANIC COMPOUNDS

Isomerism is the existence of different compounds with the same molecular formula. It includes
- structural isomerism, e.g. the position of substituents in a chain or in a benzene ring. **25.3, 25.4.1 25.9.1**
- **stereoisomerism:**
geometrical, e.g. *cis–trans* isomerism of alkenes such as RCH=CHR. **27.4.3, 25.9.2**

Members of a **homologous series** have the same functional group and differ in molecular formula by CH_2, e.g. alkanes, **25.2–5**
alkenes **25.4.1**

Nature of bonds
Alkanes are saturated; they have σ bonds between carbon atoms. **5.2.4, 25.1**
Alkenes are unsaturated; they have a C=C double bond which consists of a σ bond and a π **bond**. The π electrons are accessible to attack by electrophiles, e.g. HBr. **5.2.7, 27.4.3, 27.5**

Shapes of molecules
ethane **5.2.4, 25.1**
ethene **5.2.7, 27.4.3, 27.5**

FOUNDATION

Fractional distillation of petroleum gives hydrocarbon fractions with different boiling temperature ranges.
26.1.1, 26.1.2

Cracking converts heavy oils into more valuable alkanes, for use as fuels, and alkenes, for use in the synthesis of other compounds. **26.3.2–4**

Reforming converts the fragments from cracking into branched-chain and ring hydrocarbons. **26.3.4**

Alkanes are hydrocarbons of formula C_nH_{2n+2}. They are unreactive towards many reagents. They are used as fuels in e.g. the internal combustion engine.
26.3.1

Alkanes undergo **substitution reactions**, e.g.
$CH_4(g) + Cl_2(g) \rightarrow CH_3Cl(g) + HCl(g)$
The reaction takes place in sunlight by a **free radical** mechanism. **26.3.8**

HYDROCARBONS

Combustion of hydrocarbon fuels produces CO_2, H_2O and also the pollutants SO_2, NO_x, CO and unburnt hydrocarbons. **26.3.1, 26.4**
- CO_2 and hydrocarbons contribute to the greenhouse effect. **23.9**
- Hydrocarbons contribute to smog. **27.11**
- SO_2 and NO_x contribute to acid rain. **21.14, 26.4, 27.11**
- Addition of lead compounds to petrol is being phased out. **26.3.1, 26.4**

Alkenes C_nH_{2n} have a C=C double bond which is attacked by electrophiles in **addition reactions**.
+ hydrogen → alkane **27.7.1**
+ halogen → dihalogenoalkane **27.7.3**
+ hydrogen halide → halogenoalkane **27.7.4**
+ steam with catalyst → alcohol **27.7.7**
+ cold, dilute manganate(VII) → diol
polymerisation → poly(alkene) **27.7.10**
The double bond is **ruptured** by ozonolysis and by hot, conc. manganate(VII). **27.7.9**

MODULE 4820: FOUNDATION

1. Atoms, molecules and stoichiometry

(a) relative atomic, isotopic, molecular and formula masses	3.3–4
(b) the mole; the Avogadro constant	3.5
(c) mass spectra	1.8
(d) calculation of isotopic mass, A_r	1.8.1
(e) definition of empirical and molecular formulae	3.7, 3.8
(f) calculation of empirical and molecular formulae	3.7, 3.8
(g) calculation from equations of:	
(i) reacting masses	3.10
(ii) volumes of gases	3.11, 3.12
(iii) volumes and concentrations of solutions	3.13, 3.14
(h) stoichiometric relationships	3.14

See **Additional Topic Foundation 1** for further practice.

2. Atomic structure

(a) protons, neutrons and electrons	1.3–6, 1.9.2
(b) the distribution of mass and charges in an atom	1.4
(c) proton and nucleon numbers	1.6, 1.7
(d) isotopes	1.7
(e) 1s, 2s, 2p, 3s, 3p, 3d, 4s and 4p orbitals	2.4
(f) the shapes of s and p orbitals	2.3.3
(g) the electronic configuration of atoms and ions	2.4, 2.7
(h) the term 'first ionisation energy'; factors influencing ionisation energy, related to reactivity	2.2.2, 15.1, 15.2
(i) successive ionisation energy data and the electronic configurations of elements	2.4–6
(j) successive ionisation energy data of an element related to its position in the Periodic Table	2.2.2, 15.1, 15.2

3. Chemical bonding

(a) ionic (electrovalent) bonding	4.2, 4.2.1, 4.2.4
(b) covalent bonding and coordinate bonding	4.3, 4.6
(c) the shapes of molecules explained by electron-pair repulsion	5.1
(d) covalent bonding in σ and π bonds	5.2
(e) the shapes of ethane and ethene molecules in terms of σ and π bonds	5.1–3
(f) shapes of and bond angles in molecules mentioned in (c) and (e)	
(g) bond energy	10.5, 10.6
bond length	4.5.2
bond polarity	4.5.3
(h) hydrogen bonding	4.7.3
(i) intermolecular forces	4.7.1, 4.7.2
(j) kinetic-molecular model of melting, the liquid state and vaporisation	7.1, 8.1, 8.2
(k) metallic bonding	6.2.1
(l)(m)(n) physical properties of substances related to:	
ionic bond	4.4, 6.3
covalent bond	4.3.2, 4.4, 6.4–8
hydrogen bond	4.7.3
intermolecular forces	4.7.1, 4.7.2
metallic bond	6.2.1

See **Additional Topic Foundation 2** for further practice.

(o) energy transfers associated with the breaking and making of chemical bonds	10.1, 10.2, 10.4

4. Chemical energetics

(a) energy changes in chemical reactions; exothermic and endothermic	10.1.1
(b) (i) enthalpy change of reaction, activation energy and standard conditions	10.2–4
(ii) bond energy	10.5, 10.6
(c) enthalpy changes from experimental results	10.3
(d) Hess' Law used in calculations on determining enthalpy changes	10.4.1
(e) reaction pathway diagrams, showing enthalpy change of reaction and activation energy	14.9.2, 14.10

FOUNDATION

5. The Periodic Table: chemical periodicity

Note that UCLES students need study only the third period members of Groups 1, 3, 5 and 6; that is sodium, aluminium, phosphorus and chlorine. The study should be in sufficient detail to enable the student to appreciate the trends in properties across Period 3.

A-level Chemistry *covers all the groups in the Periodic Table. To cover the material which you will miss by not studying Groups 1, 3, 5 and 6,* **Additional Topic Foundation 3** *covers trends across Period 3*

For the third period,

(a) the periodicity in the variations in atomic radius, melting point and electrical conductivity of the elements in Period 3	15.1, 15.2
(b)(c)(d) explain the variations in (a) in terms of structure and bonding	15.1, 15.2 and see **Additional Topic Foundation 3**
(e) explain the term 'oxidation state'	3.16
(f) (i) the reactions of the elements of Period 3 with oxygen, chlorine and water	
(ii) the periodic variation of the formulae of the main oxides and chlorides	15.3 and see **Additional Topic Foundation 3**
(g) the reactions with water of the elements, oxides and chlorides of Period 3; interpretation of in terms of structure and bonding	15.3.1, 15.3.4 and **Additional Topic Foundation 3**
(h) (i) reaction with water of Group 1 elements Li, Na, K; interpretation in terms of ionisation energies	18.1, 18.4.2
(i) transition elements	24.1, 24.2.2
(j) the electronic configurations of iron and copper and their ions	24.1
(k) two transition elements, chosen from titanium to copper, to illustrate:	
(i) variable oxidation state	24.2.2, 24.6
(ii) coloured ions	24.13.2–9
(iii) catalytic behaviour	14.1.5, 24.7
(l) (i) redox processes in terms of electron transfer and change in oxidation state	3.15
(ii) Fe^{3+}/Fe^{2+}, MnO_4^-/Mn^{2+} and $Cr_2O_7^{2-}/Cr^{3+}$ as redox systems	3.15.1, 24.10.1, 24.10.2
(m) the terms: 'complex ion' and 'ligand'	24.13
(n) reactions of transition elements with ligands to form complexes	12.7.7, 24.13, 24.13.1, 24.13.2, 24.13.9

6. Carbon compounds: an introduction to organic chemistry

(a) nomenclature and formulae of alkanes, alkenes, arenes	25.2–6
(b) the following terms associated with alkanes and alkenes:	
(i) saturated and unsaturated hydrocarbons	5.2.4, 5.2.7, 25.1, 27.4.3, 27.5
(ii) homologous series	25.2
(iii) functional group	25.7
(iv) homolytic and heterolytic fission	25.8.2
(v) free radical substitution: initiation, propagation, termination	25.8.3
(vi) electrophile	25.7, 25.8.1–3
(vii) electrophilic addition, substitution	25.8.1, 25.8.3
(viii) isomerism, structural, stereochemical, cis-trans	25.3, 25.4.1, 25.9.1, 25.9.2, 27.4.3
(c) the shapes of the ethane and ethene molecules, in terms of σ and π bonds	5.2.7, 25.1, 27.4.3, 27.5
(d) (i) (iii) structural isomerism in alkenes	25.3, 25.4.1, 25.9.1, 25.9.2
(ii) (iv) cis–trans isomerism in alkenes	27.4.3
(e) crude oil as a source of hydrocarbons	26.1.1, 26.1.2, 26.3.2–5
(f) cracking and reforming	26.3.2–4
(g) combustion of alkanes and their use as fuels	26.1.1, 26.3.1, 26.4
(h) desirable characteristics of fuels	26.3.1
(i) the value to society of fossil fuels, their non-renewable nature, possible alternatives, the role of scientists in developing new fuels	26.1, 26.3.1, 26.4
(j) unleaded petrols	26.4
(k) environmental consequences of carbon monoxide, oxides of nitrogen and unburnt hydrocarbons from the internal combustion engine and their catalytic removal	26.4
(l) general unreactivity of alkanes	26.3.6
(m) the chemistry of alkenes as in:	
(i) combustion	27.6

(ii) substitution by chlorine and by bromine	26.3.7	similar to ozonolysis	27.7.9
(n) the mechanism of free radical substitution at methyl groups	26.3.8	(iv) polymerisation	27.7.5
(o) the chemistry of alkenes as in:		(p) electrophilic addition between bromine and alkenes	27.7.5
(i) addition of hydrogen, steam, hydrogen halides and halogens	27.7.1, 27.7.7, 27.7.3	(q) addition polymerisation, e.g. poly(ethene), poly(chloroethene)	27.7.10
(ii) oxidation by cold, dilute KMnO₄ to form a diol	27.7.8	(r) the work of chemists in research, development and production of organic chemicals	throughout **Part 4**
(iii) oxidation by hot concentrated KMnO₄ with rupture of C−C bond:			

ADDITIONAL TOPIC FOUNDATION 1: PRACTICE ON STOICHIOMETRIC RELATIONSHIPS

1. In a titration, 35.0 cm^3 of 0.150 mol dm^{-3} potassium hydroxide solution react with 40.0 cm^3 of phosphoric acid, H$_3$PO$_4$, of concentration 0.0656 mol dm^{-3}. Calculate the amounts (moles) of potassium hydroxide and phosphoric acid involved and write the equation for the reaction.

2. In an addition reaction, 5.4 g of an unsaturated hydrocarbon **X** of molar mass 54 g mol^{-1} react with 32 g of bromine.
(a) Calculate the amounts (moles) of **X** and Br$_2$ used in the reaction and hence the ratio by moles in which **X** and Br$_2$ react.
(b) Deduce the number of double bonds in one molecule of **X** and suggest its formula.

3. In a reaction between silver and nitric acid, the equation is

$$a\text{Ag(s)} + b\text{HNO}_3\text{(aq)} \rightarrow c\text{AgNO}_3\text{(aq)} + d\text{NO(g)} + e\text{H}_2\text{O(l)}$$

A mass 16.20 g of silver reacts with 33.3 cm^3 of 6.00 mol dm^{-3} nitric acid to form 17.0 g of silver nitrate and 1200 cm^3 at rtp of nitrogen monoxide. Work out the values of a, b, c, d and e in the equation.

4. A sample of pure iron of mass 0.0700 g was dissolved in dilute nitric acid, reduced to Fe^{2+} and titrated against a 0.0200 mol dm^{-3} solution of potassium manganate(VII), KMnO$_4$. The volume required was 12.50 cm^3.
(a) Calculate the amounts of Fe^{2+} and MnO$_4^-$ involved.
(b) Write the equation for the reaction between Fe^{2+} and MnO$_4^-$.

5. Manganese(II) chloride, MnCl$_2$, is prepared by a reaction between manganese(IV) oxide, MnO$_2$, and 12.0 mol dm^{-3} hydrochloric acid. A mass 8.70 g of manganese(IV) oxide reacts with 33.3 cm^3 of 12.0 mol dm^{-3} hydrochloric acid to form manganese(II) chloride.
(a) Calculate the amounts of MnO$_2$ and HCl used.
(b) Write the equation for the reaction.
(c) Say what gaseous product is formed and give its volume at rtp.

6. Cerium(IV) sulphate solution of concentration 0.100 mol dm^{-3} is titrated against a solution of sodium ethanedicarboxylate Na$_2$C$_2$O$_4$. The ethanedicarboxylate ion is oxidised to carbon dioxide.
(a) What is the reduction product of cerium(IV)?
(b) Write the equation for the reaction.
(c) What volume at rtp of carbon dioxide is evolved?

ADDITIONAL TOPIC FOUNDATION 2: PRACTICE ON TYPE OF BONDING

1.

Compound	$T_b/°C$
CH_4	−161
HF	17
HCl	−84

Explain how the differences in the boiling temperatures of the three compounds are related to differences in bonding within the compounds.

2.

Compound	Melting temperature/°C
CF_4	−163
CCl_4	−23
CBr_4	77
CI_4	167

Explain the trend in melting temperature in the compounds listed.

3.

Compound	Melting temperature/°C
MgO	2850
NaF	990
NaCl	800
NaBr	750
NaI	660

Account for (a) the difference between NaF and MgO (b) the trend from NaF to NaCl to NaBr to NaI.

4. Which type of intermolecular bond is present in each of the following?
(a) HCl (b) Br_2 (c) ICl (d) HF

5. The compound $SnCl_4$ is a colourless liquid which boils at 114 °C and melts at −33 °C. The compound $SnCl_2$ melts at 246 °C.
Suggest the cause of the difference in physical properties.

6. One of the substances which are listed in order of increasing boiling temperature is out of order. Identify the substance and put it into its proper place. Give your reasons.

    N_2 O_3 F_2 Ar Cl_2

7. One of the following is a liquid at room temperature, while the others are gaseous. Which do you think is the liquid? Give your reasons.

    CH_3OH C_3H_8 N_2 CO

ADDITIONAL TOPIC FOUNDATION 3: VARIATION IN PROPERTIES OF THE ELEMENTS IN PERIOD 3 (SODIUM TO ARGON)

Periods contain elements whose character changes from metallic to non-metallic. The changing properties across a period are related to the change in the nuclear charge and size of the atoms and the increasing number of outer-shell electrons. The major properties which change across a period are

- structure and bonding,
- acid–base properties,
- redox properties,
- solubility and complexing properties.

The Periodic Table in Outline

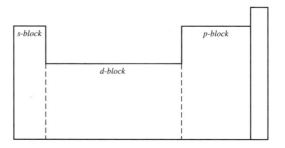

PROPERTIES OF s-BLOCK, d-BLOCK AND p-BLOCK ELEMENTS

	s-block	*d-block*	*p-block*
Structure and bonding	Form simple ionic compounds.	Oxides and chlorides are either ionic with a high degree of covalent character or covalent. The structure may be layered or macromolecular.	The oxides, hydroxides and chlorides are covalently bonded molecular compounds.
Acid–base properties	The oxides and hydroxides are basic; the chlorides are neutral.	The oxides and hydroxides are basic and insoluble with some amphotericity. The chlorides are hydrolysed to give acidic solutions.	The oxides, chlorides and hydroxides are acidic.
Redox properties	Strong reductants, e.g. the action of sodium on water.	Weak reductants, e.g. the action of iron on water.	Trend from weak reductants through weak oxidants to very strong oxidants in Group 7.
Solubility and complexing properties	Do not form complexes.	Form many complexes with a wide range of ligands. Some chlorides are insoluble.	The anions act as ligands in complex formation.

The properties of s-block, d-block and p-block elements are reviewed.

FOUNDATION

PROPERTIES OF ELEMENTS IN THE THIRD PERIOD

The third period starts with sodium in Group 1 and continues through magnesium, aluminium, silicon, phosphorus, sulphur, chlorine and argon.

The third period ...
... Na, Mg, Al, Si, P, S, Cl, Ar ...
... with the exception of the noble gas argon ...
... shows gradation in properties ...

Standard electrode potential/V

Na	Mg	Al	Si	P	S	Cl
−2.71	−2.37	−1.66	−	−	−0.48	+1.36

- Na, Mg, Al: Powerful reducing agents
- Si: Very weak reducing agent
- P: Weak reducing agents and weak oxidising agents
- S: (Weak reducing agents and weak oxidising agents)
- Cl: Powerful oxidising agent

Examples of redox reactions

The first element in the equation is acting as a reducing agent:

$2Na(s) + 2H_2O(l) \rightarrow H_2(g) + 2NaOH(aq)$

$Mg(s) + 2H^+(aq) \rightarrow Mg^{2+}(aq) + H_2(g)$

$2Al(s) + 3Cl_2(g) \rightarrow 2AlCl_3(s)$

$Si(s) + 2Cl_2(g) \rightarrow SiCl_4(l)$

$2P(s) + 5O_2(g) \rightarrow 2P_2O_5(s)$

$S(s) + Cl_2(g) \rightarrow SCl_2(l)$

... from reducing agents to oxidising agents ...

The first element in the equation is acting as an oxidising agent:

$S(s) + Cu(s) \rightarrow CuS(s)$

$Cl_2(g) + H_2(g) \rightarrow 2HCl(g)$

$S(l) + H_2(g) \rightarrow H_2S(g)$

Vigour of reaction with oxygen

... in reaction with oxygen ...

Na Mg Al	Si	P S	Cl Ar
React vigorously	Reacts slowly	React vigorously	Do not react

Electronegativity

... in electronegativity ...

Na	Mg	Al	Si	P	S	Cl
0.9	1.2	1.5	1.8	2.1	2.5	3.0

Compound formation

... in the type of bonding ...

Na Mg Al	Si P	S Cl	Ar
Form cations and therefore ionic compounds	Form covalent compounds	Form ionic compounds and covalent compounds	Does not form stable compounds

Oxidation states in hydrides, oxides and chlorides

... in oxidation states ...

Na	Mg	Al	Si	P	S	Cl	Ar
+1	+2	+3	+4 in oxides and chlorides	+3, +5 in oxides and chlorides	+6 in oxides, +4 in oxides and chlorides +2, +1 in chlorides	+7, +5, +3, +1 in oxides −1 in HCl	0

(Na, Mg, Al: +1, +2, +3 in hydrides, oxides and chlorides)

Reaction with hydrogen

... and reaction with hydrogen.

Na Mg	Al Si P	S Cl	Ar
React vigorously when heated to form ionic, basic hydrides	Do not react	React when heated to form covalent, acidic hydrides	Does not react

Summary of reactions of the elements of Period 3

The reactions of the elements with chlorine, oxygen, hydrogen and cold dilute sulphuric acid are tabulated.

Element	Heat in dry chlorine	Heat in dry oxygen	Heat in dry hydrogen	Add cold dilute sulphuric acid
Na	Very vigorous reaction gives Na^+Cl^-	Very vigorous reaction gives $(Na^+)_2O^{2-}$ + $(Na^+)_2O_2^{2-}$	Very vigorous reaction gives Na^+H^-	Dangerously violent reaction gives $H_2(g)$ + $(Na^+)_2SO_4^{2-}$
Mg	Vigorous reaction gives $Mg^{2+}(Cl^-)_2$	Very vigorous reaction gives $Mg^{2+}O^{2-}$	Vigorous reaction gives $Mg^{2+}(H^-)_2$	Very vigorous reaction gives $H_2(g)$ + $Mg^{2+}SO_4^{2-}$
Al	Vigorous reaction gives Al_2Cl_6	Vigorous reaction gives $(Al^{3+})_2(O^{2-})_3$	No reaction	Vigorous reaction, after oxide layer is removed, gives $H_2(g)$ + $(Al^{3+})_2(SO_4^{2-})_3$
Si	Slow reaction gives $SiCl_4$	Slow reaction gives SiO_2	No reaction	No reaction with dilute acid
P	Slow reaction gives PCl_3, PCl_5	Vigorous reaction gives P_2O_3, P_2O_5	No reaction	No reaction with dilute acid
S	Slow reaction gives SCl_2, S_2Cl_2	Slow reaction gives SO_2	Very slow reaction gives H_2S	No reaction
Cl	No reaction	No reaction	Vigorous reaction gives HCl	No reaction
Ar	No reaction	No reaction	No reaction	No reaction

FOUNDATION

TRENDS IN THE COMPOUNDS OF PERIOD 3

Chlorides

No chloride of argon exists. The chlorides of the second short period are shown in the table.

Chlorides	NaCl	MgCl$_2$	AlCl$_3$	SiCl$_4$	PCl$_3$, PCl$_5$	SCl$_2$, S$_2$Cl$_2$	Cl$_2$
State at room temperature	s	s	s	l	l	l	l	g
Bonding	Ionic	←—Ionic with some covalent character—→		←———————— Covalent ————————→			
Structure	Three-dimensional ionic structure	←— Layer structure —→		←———————— Molecular ————————→			
Reaction with water	Dissolves to give aqua ions, Na$^+$(aq) + Cl$^-$(aq)	Dissolves to give aqua ions, Mg^{2+}(aq) + 2Cl$^-$(aq)	Hydrolyses to give Al(OH)$_3$(s) + 3H$^+$(aq) + 3Cl$^-$(aq)	Hydrolyses to give SiO$_2$(s) + 4H$^+$(aq) + 4Cl$^-$(aq)	Hydrolyses to give H$_3$PO$_3$(aq) + 3H$^+$(aq) + 3Cl$^-$(aq)	Hydrolyses to give S(s) + H$^+$(aq) + Cl$^-$(aq)	Partially hydrolyses to give HClO(aq) + H$^+$(aq) + Cl$^-$(aq)

Oxides

No oxide of argon exists. The oxides of the second short period are shown in the table.

Oxides	Na$_2$O	MgO	Al$_2$O$_3$	SiO$_2$	P$_2$O$_5$	SO$_2$, SO$_3$	Cl$_2$O$_7$
Structure	Ionic	Ionic	Ionic with covalent character	Macromolecular covalent		Molecular covalent	Molecular covalent
Acid–base character	Basic	Basic	Amphoteric	←———————— Acidic ————————→			

The compounds of Period 3, Na to Cl, are reviewed. The chlorides of Period 3 show gradation ...
... from solid through liquid to gas ...
... from ionic to covalent bonding ...
... from dissolving in water to being hydrolysed by water.
The oxides show gradation ...
... from ionic and basic to covalent and acidic.

Reaction with water

e.g. Na$_2$O(s) + H$_2$O(l) → 2NaOH(aq)
     sodium oxide   sodium hydroxide

SO$_3$(g) + H$_2$O(l) → H$_2$SO$_4$(aq)
sulphur(VI) oxide (sulphur trioxide)   sulphuric acid

SO$_2$(g) + H$_2$O(l) → H$_2$SO$_3$(aq)
sulphur(IV) oxide (sulphur dioxide)   sulphurous acid

Cl$_2$O$_5$(g) + H$_2$O(l) → 2HClO$_3$(aq)
chlorine(V) oxide   chloric(V) acid

Hydroxides

Na Mg	Al	Si P	S Cl	Ar
Ionic	Amphoteric	Acidic	Strongly acidic	None
Strongly basic				

The hydroxides show gradation ... ... from basic to acidic.

The alkali metals in Group 1 are named for their hydroxides which are strongly basic and are alkalis (soluble bases). The alkaline earths in Group 2 are named for their strongly basic hydroxides. In the p block the corresponding compounds are strongly acidic, e.g. sulphurous acid, H_2SO_3, sulphuric acid, H_2SO_4, chloric(I) acid, HClO, chloric(V) acid, $HClO_3$.

Hydrides

Na Mg	Al	Si P	S Cl	Ar
Ionic	None formed	Unstable	Covalent	None formed
Basic			Acidic	

The hydrides of Groups 1 and 2 are ionic and basic, e.g.

The hydrides show gradation ... ... from ionic and basic to covalent and acidic.

$$Na^+(s)\ H^-(s) + H_2O(l) \rightarrow Na^+(aq)\ OH^-(aq) + H_2(g)$$

The hydrides of Groups 6 and 7 are covalent and acidic, e.g.

$$HCl(g) + H_2O(l) \rightarrow H_3O^+(aq) + Cl^-(aq)$$

Other compounds

Other compounds are tabulated.

	Na	Mg	Al	Si	P	S	Cl	Ar
Carbonate	✓	✓	←——None formed——→					
Hydrogencarbonate	✓	✓	←——None formed——→					
Sulphate	✓	✓	✓	←——None formed——→				
Nitrate	✓	✓	✓	←——None formed——→				

CHECKPOINT ON ADDITIONAL TOPIC FOUNDATION 3

1. Choose from the elements: Na, Mg, Al, Si, P, S, Cl, Ar.

(a) Name an element which exists as molecules containing (i) 1 atom, (ii) 2 atoms, (iii) 4 atoms, (iv) 8 atoms.

(b) Say which elements are (i) s-block, (ii) d-block, (iii) p-block.

2. Magnesium chloride is a solid of high melting temperature, aluminium chloride is a solid which sublimes readily at 180 °C, and silicon tetrachloride is a volatile liquid. Explain how the differences in chemical bonding account for these differences in volatility.

3. Draw dot-and-cross diagrams of the outer electron shells to show the bonding in NaCl, MgS, PH_3 and $SiCl_4$.

4. Choose from the elements: Na, Mg, Al, Si, P, S, Cl, Ar.

(a) List the elements that react readily with cold water to form alkaline solutions.

(b) List the elements that have hydrides with low boiling temperatures.

(c) Give the formulae of the oxides of the elements in their highest oxidation states and the formulae of the corresponding acids or hydroxides.

(d) List the elements that form nitrates.

(e) What is the most ionic compound that can be formed by the combination of two of these elements?

(f) Which element has both metallic and non-metallic properties?

(g) List the elements that normally exist as molecules.

5. Describe the pattern in the chemistry of the chlorides of the elements from Na to S in Period 3 of the Periodic Table.

6. Contrast the chemistry of magnesium with that of sulphur. Comment on the oxidation states used by the elements, the type of bonding in their oxides, chlorides and hydrides, the acidic or basic nature of their oxides and hydroxides and the stability of their chlorides to hydrolysis.

EXAMINATION QUESTIONS ON FOUNDATION MODULE

1. Explain, using examples, the following terms.

(a) (i) Functional group.
(ii) Electrophile.
(7 marks)

(b) (i) Homolytic fission.
(ii) Cracking.
(7 marks)

(c) Saturation and unsaturation.
(4 marks)

(d) (i) Substitution reaction.
(ii) Addition reaction.
(7 marks)
[C, '93]

2. Industry uses millions of tonnes of sulphuric acid, H_2SO_4, in the U.K. each year. This is used in the manufacture of many important products such as paints, fertilisers, soap, plastics, dyestuffs and fibres. The sulphuric acid may be prepared from sulphur in a 3-stage process.

Stage 1
The sulphur is burnt in oxygen to produce sulphur dioxide, SO_2.

$$S + O_2 \rightarrow SO_2$$

Stage 2
In the presence of a catalyst, the sulphur dioxide reacts with more oxygen to form sulphur trioxide, SO_3.

$$2SO_2 + O_2 \rightarrow 2SO_3$$

Stage 3
The sulphur trioxide is dissolved in concentrated sulphuric acid to form 'oleum', $H_2S_2O_7$, which is then reacted with water.

(a) (i) Write a balanced equation for the formation of sulphuric acid from oleum.
(1 mark)

(ii) How may tonnes of sulphur are required to produce 70 tonnes of sulphuric acid? (A_r: H, 1.0; O, 16; S, 32.)
(2 marks)

(iii) Comment on the change in the oxidation state of sulphur during these three stages.
(3 marks)

(b) A 50.0 cm³ sample of sulphuric acid was diluted to 1.00 dm³. A sample of the diluted sulphuric acid was analysed by titrating with aqueous sodium hydroxide. The reaction is

$$H_2SO_4 + 2NaOH \rightarrow Na_2SO_4 + 2H_2O$$

In the titration, 25.0 cm³ of 1.00 mol dm⁻³ aqueous sodium hydroxide required 20.0 cm³ of sulphuric acid for neutralisation.
(i) Calculate how many moles of sodium hydroxide were used in the titration.
(1 mark)

(ii) Calculate the concentration of the diluted sulphuric acid.
(2 marks)

(iii) What was the concentration of the original sulphuric acid?
(1 mark)
[C, '93]

3. This question refers to the first 20 elements in the Periodic Table. You should refer to a copy of the Periodic Table. All descriptions apply to room temperature and pressure.

From the first 20 elements, those of proton (atomic) numbers 1–20, give the symbol for:

(a) an element existing as free atoms.
(1 mark)

(b) an element that forms a chloride of formula **X**Cl which dissolves readily in water to form a neutral solution.
(1 mark)

(c) the element with the highest first ionisation energy.
(1 mark)

(d) an element that forms an oxide with a giant molecular structure.
(1 mark)

(e) an element that forms two acidic oxides of formulae **X**O_2 and **X**O_3.
(1 mark)

(f) an element that forms a liquid chloride which has a tetrahedral molecule and which reacts with water.
(1 mark)

(g) an element that forms two chlorides in which the oxidation states of the element are $+3$ and $+5$ respectively.
(1 mark)
[C, '93]

4. A sample of oxygen consisting mainly of the isotope oxygen-16 was enriched with oxygen-18. The composition of the mixture was 75.0% oxygen-16 and 25.0% oxygen-18, by volume.
The oxygen sample reacted with sodium as follows:

$$4Na(s) + O_2(g) \rightarrow 2Na_2O(s)$$

(a) Complete the table below to show the composition of some of the species involved in the reaction. (A_r: ^{16}O, 16.0; ^{18}O, 18.0; ^{23}Na, 23.0)

Species	Protons	Neutrons	Electrons
$^{23}_{11}Na$			
$^{16}_{8}O$			
$^{18}_{8}O^{2-}$			

(4 marks)

(b) Write down the electronic configuration of:
(i) a sodium atom;
(ii) an oxide ion.
(2 marks)

(c) State **three** physical properties of sodium oxide.
(2 marks)

(d) Calculate the relative atomic mass of oxygen in the sample above.
(2 marks)

(e) Some carbon-12 was burned in another sample of the oxygen mixture. The carbon dioxide produced gave the following mass spectrum.

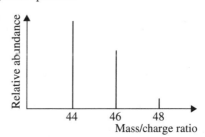

Identify the peaks shown.
(2 marks)
[C, '93]

5. In recent years, the age of volcanic rocks has been calculated by the analysis of helium isotopes. A typical sample of helium exists as its helium-4 isotope and extremely small proportions of helium-3. Analysis has shown that volcanic rocks have a higher proportion of helium-3 than a typical sample of helium has.
A sample of helium from a volcanic rock was found to have the following percentage composition, by mass:

^{3}He, 0.992%; ^{4}He, 99.008%.

(a) Explain the term *isotope*.
(1 mark)

(b) State the difference between the atomic structures of helium-3 and helium-4.
(1 mark)

(c) Relative atomic masses, A_r, can be used to compare the masses of atoms of different elements.
(i) What isotope is used as the standard for relative atomic mass measurements?
(1 mark)
(ii) Calculate the relative atomic mass of the volcanic helium sample above.
(2 marks)

(d) Helium has the largest first ionisation energy of all of the elements.
(i) Define the term *first ionisation energy*.
(2 marks)
(ii) Write an equation to represent the first ionisation energy of helium.
(1 mark)
(iii) Suggest why helium has the largest first ionisation energy of all of the elements.
(2 marks)
[C, '95]

6. Using **one** example in each case and with the aid of diagrams, describe the formation of ionic, covalent and dative covalent bonds.
(12 marks)

(b) Describe how the principle of *electron pair repulsion* helps to explain the shapes of simple molecules. Using diagrams, illustrate how this determines the molecular shapes and bond angles in the following molecules.
(i) CH_4
(ii) NH_3
(iii) SF_6
(iv) SF_2
(13 marks)
[C, '95]

7. Nitrogen compounds have a range of important uses. Two important examples are the fertilizer ammonium nitrate, NH_4NO_3, and the anaesthetic dinitrogen oxide, N_2O.

(a) The fertilizer, ammonium nitrate, NH_4NO_3, can be prepared by the neutralisation of aqueous ammonia with nitric acid:

$$NH_3(aq) + HNO_3(aq) \rightarrow NH_4NO_3(aq)$$

State the oxidation states of nitrogen in
(i) NH_3, (ii) HNO_3.
(2 marks)

(b) A 25.0 cm^3 sample of 2.00 mol dm^{-3} aqueous ammonia was neutralised by 0.500 mol dm^{-3} nitric acid.
(i) Calculate the volume of nitric acid that was used.
(2 marks)
(ii) How could a solid sample of ammonium nitrate be obtained from the reaction mixture?
(1 mark)

(c) The anaesthetic dinitrogen oxide, N_2O, can be prepared by the thermal decomposition of ammonium nitrate, NH_4NO_3:

$$NH_4NO_3(s) \rightarrow N_2O(g) + 2H_2O(g)$$

An 8.0 g sample of ammonium nitrate was heated in a carefully controlled experiment.
(i) How many moles of ammonium nitrate were used?
[A_r: N, 14; H, 1; O, 16.]
(2 marks)
(ii) Calculate the volume of dinitrogen oxide that was formed. [Assume that 1 mole of a gas occupies 24 dm^3 under the experimental conditions.]
(1 mark)
(iii) The dinitrogen oxide can be further decomposed by heat into its elements. Write an equation for this reaction.
(1 mark)
[C, '95]

8. Silicon forms compounds with hydrogen called 'silanes', similar in structure to the saturated hydrocarbons, the alkanes. Silanes form a homologous series of which the first member has the formula SiH_4.
(a) Explain the term *saturated hydrocarbon*.
(2 marks)

(b) Suggest the general formula for a silane.
(1 mark)

(c) Explain the term *structural isomer*.
(2 marks)

(d) Draw the displayed formula of each of two possible isomers of the silane containing four silicon atoms.
(2 marks)

(e) For a molecule of the silane SiH_4,
(i) draw a 'dot-and-cross' diagram showing outer electrons only,
(1 mark)
(ii) draw a diagram showing the shape and bond angles.
(2 marks)
[C, '95]

9. (a) Oil can be manufactured into many useful compounds. The following sequence describes a typical synthetic route carried out by an oil company.

The non-branched compound **A**, $C_{10}H_{22}$, was heated strongly in the absence of air to produce the saturated compound **B** and the unsaturated hydrocarbon **C** (M_r, 28). Compound **B** was reacted with bromine in UV light to form a monobrominated compound **D** and an acidic gas **E**. Compound **C** reacted with bromine in the absence of UV light to form a dibrominated product **F**.
(i) Use this evidence to identify possible structures for compounds **A** to **F** and to write any equations in the reaction sequence.
(ii) Explain why it is desirable for oil companies to 'reform' compounds such as **A** and **B**.
(13 marks)

(b) Explain the process of addition polymerisation to show the formation of **two** named polymers and comment on the problems that are associated with their disposal.
(6 marks)

(c) The hydrocarbon benzene is relatively inert to electrophilic addition. In the light of this fact, comment on the two possible structures of benzene below:

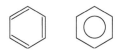

(6 marks)
[C, '95]

10. Ethanol, C_2H_5OH, is one of the most important industrial organic chemicals and is used as a fuel, solvent, antifreeze, and intermediate in the synthesis of many organic chemicals.

(a) State **two** desirable characteristics of a fuel.
(2 marks)

(b) Ethanol can be prepared industrially by the reaction of ethene and steam:

$$C_2H_4(g) + H_2O(g) \rightarrow C_2H_5OH(g)$$

The standard enthalpy change for this reaction can be determined by using the standard enthalpy changes of formation given in the table.

$\Delta H_f^\ominus[C_2H_4(g)]$/kJ mol^{-1}	$\Delta H_f^\ominus[H_2O(g)]$/kJ mol^{-1}	$\Delta H_f^\ominus[C_2H_5OH(g)]$/kJ mol^{-1}
+52.3	−241.8	−277.7

(i) Define the term *standard enthalpy change of formation*.
(2 marks)
(ii) Calculate the standard enthalpy change for this reaction.
(3 marks)

(c) Ethanol dissolves readily in water because of hydrogen bonding between molecules of ethanol and water.
(i) Explain the term *hydrogen bonding*.
(2 marks)
(ii) Draw a diagram, including any relevant dipoles, to show the hydrogen bonding between a molecule of ethanol and water.
(2 marks)
[C, '95]

11. Antimony, symbol Sb, proton (atomic) number 51, has been known since about 4000 BC. Nowadays, its main use is to harden and to strengthen lead alloys.

(a) A typical sample of antimony consists of two isotopes and has the following composition, by mass: ^{121}Sb, 57.25%; ^{123}Sb, 42.75%.

(i) How could this information be obtained experimentally?
(1 mark)
(ii) What can be deduced about the atomic structure of these two isotopes?
(2 marks)
(iii) Calculate the relative atomic mass of the antimony sample.
(2 marks)

(b) Antimony is produced in a two-stage process from the sulphide ore, Sb_2S_3. The ore is first roasted in oxygen to form the oxide.

$$2Sb_2S_3(s) + 9O_2(g) \rightarrow Sb_4O_6(s) + 6SO_2(g)$$

The oxide is then reduced with carbon.

$$Sb_4O_6(s) + 3C(s) \rightarrow 4Sb(s) + 3CO_2(g)$$

(i) State the oxidation state of antimony in its oxide.
(1 mark)
(ii) Showing each stage of your working clearly, calculate the volume of carbon dioxide that would be produced by the processing of 10 moles of Sb_2S_3. (Assume that 1 mole of a gas occupies 24 dm^3 under the experimental conditions.)
(3 marks)
[C, '95]

12. This question concerns the following oxides: Na_2O; MgO; SiO_2; SO_3

(a) Showing outer electrons only, draw a dot-and-cross diagram for the oxide MgO.
(2 marks)

(b) From the list above, identify the oxide that best fits the description given.
(i) An oxide that reacts with water forming a strongly alkaline solution.
(1 mark)
(ii) An oxide that is insoluble in water.
(1 mark)
(iii) An oxide that is slightly soluble in water forming a weakly alkaline solution.
(1 mark)
(iv) An oxide that reacts vigorously with water forming a strongly acidic solution.
(1 mark)
(v) An oxide that has a simple molecular structure at room temperature.
(1 mark)

(c) From your answers to (b) above, suggest an equation for the reaction of an oxide with water to form
(i) an acidic solution,
(1 mark)
(ii) an alkaline solution.
(1 mark)
[C, '95]

4821: CHAINS AND RINGS

1. BASIC CONCEPTS

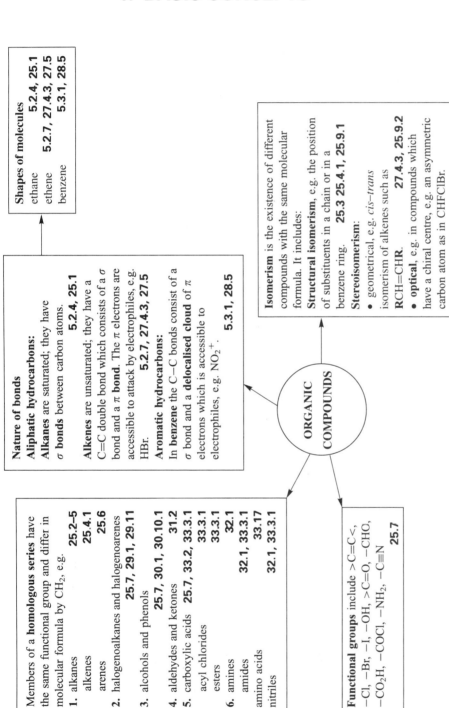

2. HYDROCARBONS

Alkanes
Alkenes
See Foundation 6. Carbon compounds: an introduction to Organic Chemistry.

Arenes contain a benzene ring. Benzene C_6H_6 has C–C bonds intermediate between C–C and C=C. The electrons in the π bonds are delocalised (**28.7.2, 28.7.7**) and are subject to attack by **electrophilic reagents**, e.g. nitration by $HNO_3 + H_2SO_4$ **28.7.1**
mechanism **28.7.2**
halogenation by e.g. $Br_2 + AlBr_3$ **28.7.6, 28.7.7**
alkylation by e.g. $RCl + AlCl_3$ **28.7.6**
acylation by e.g. $RCOCl + AlCl_3$ **28.7.6**
Benzene adds C_2 in UV light to form $C_6H_6Cl_6$. **28.6**

Methylbenzene, $C_6H_5CH_3$
1. The side chain is substituted by e.g. Cl_2 in a free radical reaction. **28.11.1**
2. The ring is substituted by electrophiles e.g. Cl_2, NO_2^+. **28.7.6, 28.7.7, 28.10**
Methylbenzene is more reactive than benzene. Substitution occurs in the 2- and 4-positions. **28.10, 28.12.3**
3. The side chain can be oxidised to $-CO_2H$. **28.11.2**

3. HALOGEN DERIVATIVES

The **order of reactivity of halogenoalkanes** is
$R_3CX > R_2CHX > RCH_2X$ **29.7, 29.8**
$RI > RBr > RCl$ **29.7**
Fluoroalkanes and fluorohalogenoalkanes, e.g. CFCs, are very unreactive because of the high strength of the C–F bond and find special uses. **29.5**
There is damage to the ozone layer from CFCs. **29.9**

Halogenoalkanes
With alkali, hydrolysis or elimination can occur:
+ OH^- (aq) → alcohols, **ROH**
+ OH^- (ethanolic) → alkenes
Other reactions include:
+ CN^- (ethanolic) → nitriles, **RCN**
+ NH_3 (ethanolic) → amines, **RNH_2**. **29.7**

The **mechanisms** of these reactions depend on the polar $C^{\delta+}$–Halogen$^{\delta-}$ bond which is attacked by nucleophiles, e.g. OH^-. **29.8**

Halogenoarenes are much less reactive than halogenoalkanes because the C–X bond interacts with the ring. **29.13**

4. HYDROXY COMPOUNDS

Alcohols: include:
primary alcohols, 1°, RCH_2OH
secondary alcohols, 2°, R_2CHOH
tertiary alcohols, 3°, R_3COH
30.1.1, 30.1.2

They can be distinguished by the ease of oxidation and Lucas' test. **30.7.2**

Reactions of alcohols
1. Oxidation by e.g. $Cr_2O_7^{2-}/H^+$
   1°, RCH_2OH oxidised to aldehydes $RCHO$ and then to carboxylic acids RCO_2H
   2°, R_2CHOH oxidised to ketones, R_2CO
   3°, R_3COH oxidised with difficulty. **30.7.2**
2. Combustion $\to CO_2 + H_2O$ **30.5**
3. Reaction with $Na \to RONa + H_2$ **30.7.1**
4. Esterification by acids or acid chlorides to esters RCO_2R'. **30.7.1**
5. Reaction with HCl or $P + I_2$ or PBr_3 or PCl_5 or $SOCl_2 \to$ halogenoalkanes. **30.7.2**
6. Dehydration by conc H_2SO_4 to alkenes. **30.7.2**

Ethanol, C_2H_5OH
The fermentation method is used to make alcoholic drinks. **30.3.3**
The hydration of ethene is used to make ethanol for industrial use. **30.4.3**

The **iodoform test** for $CH_3CH(OH)-$ compounds is the formation of tri-iodomethane with $I_2 + NaOH(aq)$. **31.6.7**

Phenol, C_6H_5OH, is a weak acid, stronger than C_2H_5OH and weaker than CH_3CO_2H. **30.12.1**
- The $-OH$ group reacts with NaOH or $NaHCO_3$ to give C_6H_5ONa. **30.12.1**
- Phenol is more reactive than benzene to electrophiles, e.g. Br_2, NO_2^+, which substitute in the ring to give $C_6H_2Br_3OH$ and $C_6H_2(NO_2)_3OH$. **30.12.2**

5. CARBONYL COMPOUNDS

Aldehydes, $RCHO$, can be made by oxidation of primary alcohols, RCH_2OH. **Ketones** can be made by the oxidation of secondary alcohols, R_2CHOH, by e.g. acidified dichromate. **31.4.1-3, 31.5**

Aldehydes, $RCHO$, are reduced to primary alcohols by e.g. $NaBH_4$. **31.6.1**
Ketones, R_2CO, are reduced to secondary alcohols by e.g. $NaBH_4$. **31.6.1**

Oxidation
Aldehydes are oxidised to carboxylic acids by e.g. acidified dichromate; ketones are difficult to oxidise. **31.6.2**
Aldehydes (not ketones) **reduce**
- Tollens' reagent, $[Ag(NH_3)_2]^+(aq)$ to Ag,
- Benedict's solution and Fehling's solution, Cu^{2+} complex ions, to Cu_2O. **31.6.3**

Common to **aldehydes and ketones** are:
- **addition reactions** with nucleophiles, e.g. $+HCN \to$ cyanhydrin, R_2COHCN **31.3, 31.3.1, 31.3.2, 31.7** mechanism. **31.7**
- addition-elimination reactions with e.g. 2,4-dinitrophenylhydrazine (Brady's reagent) give coloured crystalline solids used for identification. **31.6.5**

The **iodoform test** for $CH_3C=O$ or CH_3CHOH-: Add $I_2 + OH^-$ (aq) or $KI + KClO$; tri-iodomethane (iodoform), CHI_3, is formed. **31.6.7**

CHAINS AND RINGS

6. CARBOXYLIC ACIDS AND DERIVATIVES

Carboxylic acids, RCO_2H ionise:
$RCO_2H(aq) + H_2O(l)$
$\rightleftharpoons RCO_2^-(aq) + H_3O^+(aq)$
Delocalisation of charge stabilises the RCO_2^- ion. **33.4**
Substituents alter the strength of the acid. **12.7.7**
Carboxylic acids are made by
- oxidation of aldehydes $RCHO$ and primary alcohols RCH_2OH by e.g. $K_2Cr_2O_7$ + acid,
- hydrolysis of nitriles.
33.5.2, 33.6.1, 33.6.2, 33.6.3

Reactions of RCO_2H include:
- formation of salts, e.g. RCO_2Na. **33.8.1**
- esterification by alcohols, **33.8.2**
- formation of acyl chlorides, $RCOCl$, with $SOCl_2$ or PCl_5. **33.8.3**

Esters, RCO_2R'
Prepared from alcohol or phenol + carboxylic acid or acyl chloride. **33.8.2, 33.10.2**
Used as solvents and as food flavourings.

Additional Topic Chains and Rings 1
Hydrolysed by acids and bases. **33.13.1**
Polyesters **10.8**
Fats are natural esters, and are hydrolysed in the manufacture of soaps. **33.13.3**

Acyl chlorides, $RCOCl$
Reactivity depends on attack by nucleophiles on the $C^{\delta+}$ atom in the $C^{\delta+}=O^{\delta-}$ group which carries a higher $\delta+$ charge than C in the $\overset{\delta+}{C}-\overset{\delta-}{Cl}$ of chloroalkanes, RCl, and C in $\overset{\delta+}{C}=\overset{\delta-}{O}$ of aroyl chlorides, $ArCOCl$. **33.10.1**

Reactions of $RCOCl$ include:
- hydrolysis to $RCO_2H + HCl$, **33.10.2**
- reaction with alcohols, $R'OH$, to form esters, RCO_2R' and with phenols, as $ArONa$, to form esters RCO_2Ar,
- reaction with ammonia to form amides, $RCONH_2$ and with amines $R'NH_2$, to form amides, $RCONHR'$. **33.10.2**

7. NITROGEN COMPOUNDS

Amino acids $H_2NCHRCO_2H$ ionise as acids and as bases and form **zwitterions** $H_3\overset{+}{N}CHRCO_2^-$ (aq). **33.17**
Condensation polymerisation gives **peptides**, $H_2NCHRCONHCHRCO_2H$. Further polymerisation gives polypeptides and proteins. The hydrolysis of **proteins** yields amino acids.

33.17.2 and Additional Topic Chains and Rings 2

Reactions:
Acylation by $R'COCl$ gives an amide
$RNH_2 + R'COCl \rightarrow R'CONHR + HCl$ **33.10.2**

Amides $RCONH_2$, $RCONHR'$
- made from $RCOCl + NH_3$ or RNH_2
- hydrolysed by acids and alkalis. **33.14**

Amines include primary RNH_2, secondary R_2NH, tertiary R_3N, quaternary R_4N^+.
Amines are **basic**:
$RNH_2 + H^+ \rightleftharpoons RN\overset{+}{H}_3$
Basicity depends on the availability of a lone pair on the N atom: **32.6**
$RNH_2 > NH_3 > ArNH_2$

Aromatic amines
e.g. phenylamine, $C_6H_5NH_2$
1. with bromine → 2,4,6-tribromophenylamine, **32.8.4**
2. with $NaNO_2 + HCl(aq)$ (cold) → benzenediazonium chloride
$C_6H_5-\overset{+}{N}\equiv N\ Cl^-$ **32.8.3**
which couples with phenol to form an azo dye $C_6H_5-N=N-C_6H_4OH$. **32.10**

Preparation of amines
- from halogenoalkanes, $RX(alc) + NH_3(alc)$ **32.4.1**
- from nitriles $RCN + LiAlH_4$ or $Na/ethanol$ **32.7.2**
- from $ArNO_2 + Sn + HCl$ **32.4.2, 32.7.2**

8. POLYMERISATION

MODULE 4821: CHAINS AND RINGS

Condensation polymers include
- poly(esters), e.g. Terylene, formed by condensation polymerisation between a dicarboxylic acid and a diol, **33.13.2**
- poly(amides), e.g. nylon 6 and nylon 6,6, formed by condensation polymerisation between a diamine and a dicarboxylic acid, **33.14.1, 35.1.3**
- **peptides and proteins**, formed by condensation polymerisation of α-amino acids, $H_2NCHRCO_2H$. **33.17.2**

Condensation polymerisation
Monomers which have two functional groups react by the elimination of small molecules, e.g. H_2O or HCl, to form a polymer. **33.13.2, 33.14.1**

POLYMERS

Addition polymerisation of a monomer M to give a polymer M_n:
$nM \rightarrow M_n$
takes place by a free radical **mechanism**. **27.7.10**

Addition polymers, poly(alkenes), include: high-density poly(ethene); low-density poly(ethene); poly(propene); poly(chloroethene), PVC; and poly(phenylethene). **35.1.1, 35.1.2**

Disposal of waste plastics is a problem. Waste plastics may be dumped in landfill sites. Since most plastics are not biodegradable, they remain without decomposing for a very long time. Waste plastics may be burned and used as a fuel, but some produce harmful combustion products. Possible solutions to the problem are
- the use of **biodegradable plastics**, e.g. Biopol, which is made by micro-organisms
- plastics which slowly dissolve in water
- plastics which slowly decompose in sunlight
- **recycling of plastics** – the collection, sorting and remoulding of thermoplastics.

Additional Topic Chains and Rings 3

MODULE 4821: CHAINS AND RINGS

1. Basic concepts

(a) (i) alkanes	25.2–5
alkenes and arenes	25.4.1
(ii) halogenoalkanes and halogenoarenes	25.7, 29.1, 29.11
(iii) alcohols (primary, secondary, tertiary) and phenols	25.7, 30.1, 30.10.1
(iv) aldehydes and ketones	31.2
(v) carboxylic acids,	25.7, 33.2, 33.3.1
acyl chlorides and esters	33.3.1
(vi) amines (primary only)	32.1
amides	32.1, 33.3.1
amino acids	33.17
nitriles	32.1, 33.3.1
(b) (i) saturated and unsaturated hydrocarbons	25.4.1
(ii) aliphatic and aromatic	25.6
(iii) homologous series	25.2
(iv) functional group	25.7
(v) homolytic and heterolytic fission	25.8.2
(vi) free radical, initiation, propagation, termination	25.8.2
(vii) nucleophile, electrophile, inductive effect	25.8.3
(viii) addition, substitution, elimination, hydrolysis	25.8.1
(ix) oxidation and reduction	3.15.1
(x) isomerism: structural isomerism,	25.3, 25.4.1, 25.9.1
stereochemical: *cis–trans* isomerism,	27.4.3, 25.9.2
optical isomerism and chirality.	5.1.4, 25.9.2
(c) the shapes in terms of σ and π carbon–carbon bonds of:	
ethane,	5.2.4, 25.1
ethene,	5.2.7, 27.4.3, 27.5
benzene	5.3.1, 28.5
(d) chiral centres and optical isomerism	5.1.4, 25.9.2
(e) isomers of an organic molecule of known molecular formula	**Checkpoint 25D** and **Questions on Chapter 25**

2. Hydrocarbons

(a) the structure of benzene; delocalisation of electrons	5.3.1, 28.1
(b) the chemistry of arenes e.g. benzene and methylbenzene:	
(i) substitution reactions with chlorine and with bromine,	28.7.6, 28.7.7
(ii) nitration,	28.7.1, 28.7.2
(iii) oxidation of the side-chain to give a carboxylic acid.	28.11.2
(c) the mechanism of the electrophilic substitution in arenes, e.g. the mono-nitration of benzene	28.7.2, 28.7.7
(d) halogenation in the side-chain or in the aromatic nucleus	28.7.6, 28.7.7, 28.11.1, 28.12.3
(e) the positions of substitution in methylbenzene	28.10.1

3. Halogen derivatives

(a) reactions of halogenoalkanes: hydrolysis, formation of nitriles, formation of primary amines and the elimination of hydrogen halide	29.7
(b) the mechanism of nucleophilic substitution in primary and tertiary halogenoalkanes	29.7
(c) (e) different reactivities of primary, secondary and tertiary halogenoalkanes towards hydrolysis	29.7
(d) the different reactivities of secondary halogenoalkanes and chlorobenzene towards hydrolysis	29.13
(f) the uses of fluoroalkanes and fluorohalogenoalkanes related to their relative chemical inertness	29.5
(g) the effect of CFCs on the ozone layer	29.9

4. Hydroxy compounds

(a) the production of ethanol by	
(i) fermentation	30.3.3
(ii) reaction of steam with ethene	30.3.1
(b) the chemistry of alcohols, e.g. ethanol:	
(i) combustion,	30.5
(ii) substitution to give halogenoalkanes,	30.7.2
(iii) reaction with sodium,	30.7.1
(iv) oxidation to carbonyl compounds and carboxylic acids,	30.7.2
(v) dehydration to alkenes,	30.7.2
(vi) esterification, acylation	30.7.1
(c) primary, secondary and tertiary alcohols	30.1.1, 30.1.2
(d) reactions with $Cr_2O_7^{2-}/H^+$	30.7.2 under 'Oxidation' and Table 30.1
(e) the reactions of $CH_3CH(OH)-$ compounds to give tri-iodomethane	31.6.7
(f) reactions of phenol with bases and with sodium,	30.12.1
nitration and halogenation of the aromatic ring	30.12.2
(g) the relative acidities of water, phenol and ethanol	30.12.1

5. Carbonyl compounds

(a) (i) the formation of aldehydes and ketones by the oxidation of alcohols	31.4.1, 31.4.2, 31.4.3
(ii) their reduction to alcohols by $NaBH_4$	31.6.1
(b) the nucleophilic addition reactions of hydrogen cyanide	31.3, 31.3.1, 31.3.2, 31.7
(c) the use of 2,4-dinitrophenylhydrazine to detect carbonyl compounds	31.6.5
(d) aldehydes and ketones distinguished by Fehling's or Benedict's and Tollens' reagents	31.6.3
(e) the reaction of CH_3CO- compounds to give tri-iodomethane.	31.6.7

6. Carboxylic acids and derivatives

(a) the formation of carboxylic acids from alcohols, aldehydes and nitriles	33.5.2, 33.6.1–3
(b) the reactions of carboxylic acids to form	
(i) salts,	33.8.1
(ii) esters,	33.8.2
(iii) acyl chlorides	33.8.3
(c) the acidity of carboxylic acids,	33.4
and of chlorine-substituted ethanoic acids	12.7.7
(d) the hydrolysis of acyl chlorides	33.10.1, 33.10.2
(e) the reactions of acyl chlorides with alcohols, phenols and primary amines	33.10.2 and Figure 33.5
(f) the relative ease of hydrolysis of acyl chlorides, alkyl chlorides and aryl chlorides	33.10.1
(g) the acid- and base-catalysed hydrolysis of esters	33.13.1
(h) fats and oils as glyceryl esters	33.13.3
(i) commercial uses of esters	See **Additional Topic Chains and Rings 1**

7. Nitrogen compounds

(a) the formation of ethylamine (from a nitrile)	32.4.1
and of phenylamine (from nitrobenzene)	32.4.2, 32.7.2
(b) the relative basicities of ammonia, ethylamine and phenylamine	32.6
(c) the reactions of phenylamine with	
(i) bromine,	32.8.4
(ii) nitrous acid	32.8.3
(d) the coupling of benzenediazonium chloride and phenol	32.10, including Figure 32.6
(e) the formation of amides from $RNH_2/R'COCl$	33.10.2 and Figure 33.16
(f) acid- and base-catalysed hydrolysis of amides	33.14 and Figure 33.16
(g) the acid/base properties of amino acids; zwitterions	33.17, 33.17.1, 33.17.2
(h) the formation of peptide bonds; polypeptides	33.14.1
(i) hydrolysis of proteins	See **Additional Topic Chains and Rings 2**

8. Polymerisation		nylon 6 and nylon 66	33.14.1, 35.1.3
(a) addition polymerisation e.g. poly(phenylethene)	27.7.10, 35.1.1, 35.1.2	(c)(d)(e)(f) different types of polymerisation reaction	35.1
(b) condensation polymerisation		(g) the disposal of poly(alkene)s	See **Additional Topic Chains and Rings 3**
(i) in polyesters e.g. Terylene	33.13.2		
(ii) in polyamides e.g. peptides and proteins,	33.17.2		

ADDITIONAL TOPIC CHAINS AND RINGS 1: COMMERCIAL USES OF ESTERS

Esters are used as solvents ...

Esters are used as solvents; for example ethyl ethanoate is used as a solvent in the laboratory and in industry.

... in soap manufacture ...

Esters of glycerol are used in soap manufacture (see **33.13.3**).

... as food flavourings ...

Many esters are volatile liquids with fruity smells and fruity tastes. In the food industry, esters used as food flavourings include ethyl methanoate, $HCO_2C_2H_5$, which has the flavour of rum, propyl pentanoate, $C_5H_{11}CO_2C_3H_7$, which has the flavour of pineapple, ethyl butanoate, $CH_3(CH_2)_2CO_2C_2H_5$ which has an apple odour and octyl ethanoate, $CH_3CO_2(CH_2)_7CH_3$ which has an orange odour.

Esters used as emulsifiers include mixed esters of glycerol with fatty acids and another acid, such as ethanoic acid, 2-hydroxypropane-1,2,3-tricarboxylic acid (citric acid) or 2,3-dihydroxybutanedioic acid (tartaric acid).

Antioxidants are added to foods to retard spoilage through oxidation. They include the esters 2-butyl-4-methoxyphenol (known as BHA, with E number E 320) and 2,6-dibutyl-4-methylphenol (known as BHT, with E number E 321).

... as emulsifiers ...
... as antioxidants ...
... in the manufacture of plastics ...
... and in medicine.

Esters are used in the manufacture of plastics (see polyesters, **33.13.2**). Esters are added as plasticisers in the polymerisation of other polymers, e.g. PVC, to improve the flexibility of those plastics.

Esters used in medicine include methyl 2-hydroxybenzoate, 'oil of wintergreen', (see **33.12**) and 2-ethanoyloxybenzoic acid, aspirin (see **33.12**).

For methyl cellulose see **33.12**.

ADDITIONAL TOPIC CHAINS AND RINGS 2: THE HYDROLYSIS OF PROTEINS

The sequence of amino acids in a protein is worked out ...
... by purifying the protein ...
... hydrolysing ...
... separating the amino acids ...

The first step in working out the structure of a protein is to find out which amino acid residues it contains. The protein must first be obtained pure. It is precipitated from solution by adding a salt, e.g. ammonium sulphate, and collected by centrifugation.

Hydrolysis of the peptide bonds is carried out by heating the protein with 6 mol dm^{-3} hydrochloric acid at 100–120 °C for 10–24 hours. The amino acids are then separated, usually by chromatography, e.g. paper chromatography (see **8.7.3**), thin layer chromatography (see **8.7.4**), column chromatography, ion exchange chromatography or by electrophoresis.

Proteins have relative molecular masses of 5000 to 36 000. A molecule may contain more than one polypeptide chain, and an individual polypeptide chain contains 100–300 amino acid residues. After finding out which amino acid residues are present, the biochemist sets about finding out the order in which several hundred amino acid residues are linked by peptide bonds to form the polypeptide chain.

The sequence of amino acids in the protein chain is worked out by partially hydrolysing the protein to give small peptide fragments. The sequence of amino acids in the peptide fragments is worked out by further hydrolysis and analysis. By looking for overlaps between sequences of amino acid groups in the peptide fragments, the biochemist gradually assembles a picture of the whole chain. For example, results of an analysis might be:

Fragments obtained by hydrolysis 1 :

Asp Tyr Glu Leu Arg

His Lys

Gly Ala

Fragments obtained by hydrolysis 2 :

His Lys Asp Tyr

Glu Leu Arg Gly Ala

(Asp = aspartic acid, Tyr = tyrosine, Glu = glutamic acid, Leu = leucine, Arg = arginine, His = histidine, Lys = lysine, Gly = glycine, Ala = alanine)

... and working out the sequence by overlapping peptide fragments.

By overlapping these fragments, one can deduce the sequence:

His Lys Asp Tyr Glu Leu Arg Gly Ala.

This is the primary structure of the protein.

This is part of the long chain of amino acid residues. You can imagine the meticulous labour involved in completing the sequence for a protein consisting of several hundred amino acids. This sequence is called the primary structure of the protein.

ADDITIONAL TOPIC CHAINS AND RINGS 3: RECYCLING MATERIALS

This topic is also part of Trends and Patterns

1. RECYCLING METALS

In order to produce iron, iron ore must be mined, transported and smelted. This takes several times more energy than collecting and remelting scrap iron.

Aluminium is a prime candidate for conservation for two reasons. Owing to its very high resistance to corrosion, used aluminium is as good as new. A second reason is the high energy consumption in the extraction of aluminium from bauxite (see **19.2.2**). The cost of reusing scrap aluminium is only one twentieth of the cost of making the pure metal.

Recycling metals saves metal ores ...
... and also the energy used in extracting the metal from its ore
Iron, steel and other metals and plastics are recovered from used cars for recycling.

The collection, sorting and recycling of metals has become an important industry. A major source of used metals is motor vehicles. There are about 140 million cars in Europe and 12 million a year reach the end of the road. About 75% of this mass of about 12 million tonnes is iron and steel. The European Community has encouraged member nations to make car recycling a priority. Some German car manufacturers, e.g. BMW and Mercedes, have started to design recyclability into their vehicles. They have developed **disassembly lines** in which cars are stripped down and ferrous metals, non-ferrous metals and plastics are separated for recycling.

2. RECYCLING PLASTICS

Plastic waste constitutes about 7% of household waste. Unlike some other wastes, e.g. kitchen waste and paper, plastics are **non-biodegradable**. They are synthetic materials,

Plastic waste is non-biodegradable ...
... and uses up landfill sites.

and soil and water do not contain micro-organisms with the enzymes needed to feed on plastics and degrade them. Plastic waste is buried in **landfill sites**, and there it remains unchanged for decades. Local authorities have to find more and more landfill sites.

An alternative to dumping is **incineration**, with the possibility of making use of the heat generated. Plastics are petroleum products, and plastic waste contains about the same amount of energy as the oil from which it came. To burn plastic waste with the release of useful energy is an obvious solution to the problem. In the UK only about 10% of plastic waste is incinerated, but in some other countries, e.g. Denmark and Japan, incinerators consume over 70% of domestic waste. The plastic part of the waste assists in the incineration of other parts of domestic rubbish. If plastics are removed from domestic waste, together with paper, what is left is organic waste which is too wet to burn. If the waste is burnt with the plastics included, potentially useful energy is generated. Some plastics, however, burn with the formation of toxic gases, e.g. hydrogen chloride, from PVC, and hydrogen cyanide, from poly(propenenitrile), and incinerators must be designed to remove these gases from the exhaust.

Incineration of waste plastics releases useful energy ...
... but some plastics burn to form toxic gases.

Recycling is the most efficient use of resources. The municipal solid waste collected in the UK each year contains about 1.5 million tonnes of plastics. In addition, used plastics from large articles such as fridges, washing machines, agricultural machinery, cars etc bring the total up to about 1.8 million tonnes. A major difficulty in recycling is that, since different plastics have widely differing properties, mixed plastic waste is of limited use. While any two glass bottles can be recycled together, the same is not true of a PVC bottle and a poly(ethene) bottle. In cases where collection and separation are easy, recycling is profitable. Examples are

If plastics of different kinds can be collected separately, recycling is profitable.

- telephone hand sets of poly(ethene)
- poly(propene) car bumpers and casings from car batteries
- soft-drink bottles made of PET
- poly(styrene) insulated food containers.

Municipal solid waste contains a mixture of plastics. Some of the mixed plastics are recovered, shredded, melted and extruded in the shape of planks. This recycled plastic can be used to construct items such as agricultural fencing and garden seats. There is a limited demand for plastics of this quality.

Mixed plastic waste is of limited use for recycling.

Biodegradable plastics

Chemists have invented some biodegradable plastics.

Biopolymers In the UK, ICI markets the biopolymer poly(3-hydroxybutanoic acid), PHB, which has the trade name Biopol. It is made by certain bacteria from glucose. When Biopol is discarded, micro-organisms in the soil, in river water and in the body can break it down within nine months. At present a Biopol container is seven times the price of a poly(ethene) container. With increasing use, the price may well fall.

Biodegradable plastics may be ...
... biopolymers ...
... photodegradable plastics ...

Photodegradable plastics A Canadian firm has produced a photodegradable polymer, which they incorporate in polystyrene cups. Exposed to sunlight for 60 days, the cups break down into dust particles.

Synthetic biodegradable plastics An Italian company, Feruzzi, has produced a biodegradable polymer which is suitable for carrier bags. The material consists of poly(ethene) and up to 50% starch. When the material is buried, micro-organisms begin to feed on the starch, converting it into carbon dioxide and water, and in time the polymer chains dissolve in water. The cost at present is about twice that of a regular plastic bag.

... or synthetic biodegradable plastics.

Recycling versus biodegradability

The choice between manufacturing biodegradable plastics and recycling plastics is a difficult one.

There are two solutions to the problem of plastic waste. One is to make degradable plastics, and the other is to recycle plastics. The two solutions do not live well together. Although some biodegradable plastics can be recycled with other plastics, photodegradable plastics cannot be included. Waste plastics can be turned into items such as sacks, park benches, roofing and drain pipes. You can imagine the accidents that could occur if such materials were to break up in sunlight.

3. Recycling glass

The chemicals use in glass making are sand, limestone, sodium carbonate, metal oxides and waste glass known as cullet, which makes up about 20% of the mixture. This saves raw materials and also saves energy because cullet makes the mixture melt at a lower temperature. Bottle banks collect used bottles and jars which are then sold to glass manufacturers. The manufacturers guarantee to the local authority that operates the bottle bank that they will buy the glass at a stated price. Benefits of the scheme are:

- saving in energy used in glass manufacture;
- saving in natural resources;

Glass is easy to recycle There is a 'bottle bank' scheme for collecting waste glass.

- saving in refuse disposal costs;
- less glass litter and broken glass in the environment;
- a source of income for local authorities.

CHECKPOINT ON ADDITIONAL TOPIC CHAINS AND RINGS 3

1. Millions of plastic bags are discarded after one or two hours' use. Many plastic bags are made of poly(ethene).

(a) Explain how poly(ethene) is obtained from petroleum.

(b) How long did petroleum take to form?

(c) Can it be replaced?

(d) What is meant by the statement that plastic bags are non-biodegradable? What significance does this statement have for the disposal of plastic waste?

2. Gas is used to convert polystyrene into polystyrene foam.

(a) What is the advantage of polystyrene foam for serving food in take-away restaurants?

(b) For how long is a polystyrene foam package in use?

(c) What happens to the polystyrene it contains?

(d) What happens to the gas it contains?

3. Suggest (a) applications for a photodegradable plastic (b) items for which it would not be suitable.

4. Discuss the pros and cons of recycling plastics or using biodegradable plastics.

EXAMINATION QUESTIONS ON CHAINS AND RINGS

1. The structures of three derivatives of propane are given below.

(a) What reagents will convert

Compound A: $H-\underset{\underset{H}{|}}{\overset{\overset{H}{|}}{C}}-\underset{\underset{H}{|}}{\overset{\overset{H}{|}}{C}}-\underset{\underset{H}{|}}{\overset{\overset{H}{|}}{C}}-O-H$

Compound B: $H-\underset{\underset{H}{|}}{\overset{\overset{H}{|}}{C}}-\underset{\underset{H}{|}}{\overset{\overset{H}{|}}{C}}-\underset{\underset{H}{|}}{\overset{\overset{H}{|}}{C}}-Cl$

Compound C: $H-\underset{\underset{H}{|}}{\overset{\overset{H}{|}}{C}}-\underset{\underset{H}{|}}{\overset{\overset{H}{|}}{C}}-\underset{\underset{H}{|}}{\overset{\overset{H}{|}}{C}}-N{\overset{H}{\diagup}\underset{H}{\diagdown}}$

(i) compound **A** into compound **B**,
(ii) compound **B** into compound **C**?

(3 marks)

(b) What procedure would be employed in carrying out reaction (a)(i) to avoid loss of compound **A** to the atmosphere?
(1 mark)

(c) State the systematic name for compound **C**.
(1 mark)

(d) Draw a diagram to show how **one** of these molecules would be expected to form a hydrogen bond with a water molecule.
(2 marks)
[C, '93]

2. Citrus fruits are the source of a number of compounds used as flavours and perfumes. The following compounds are examples.

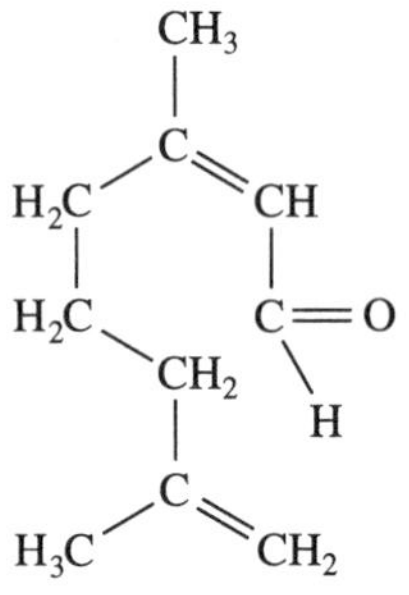

P

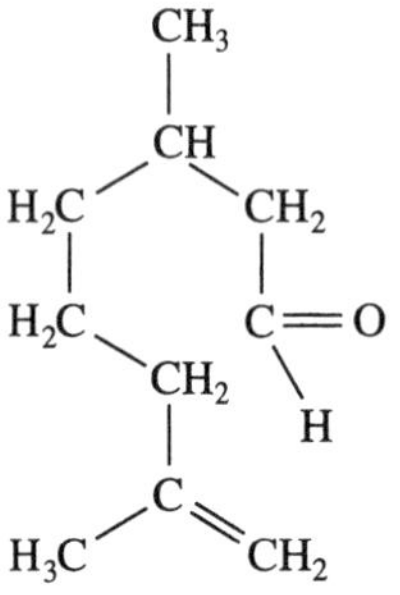

Q

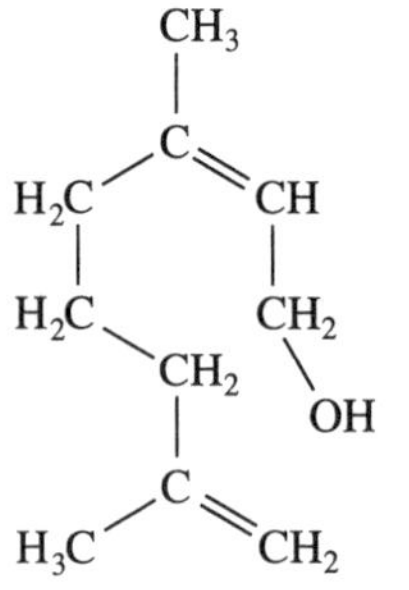

R

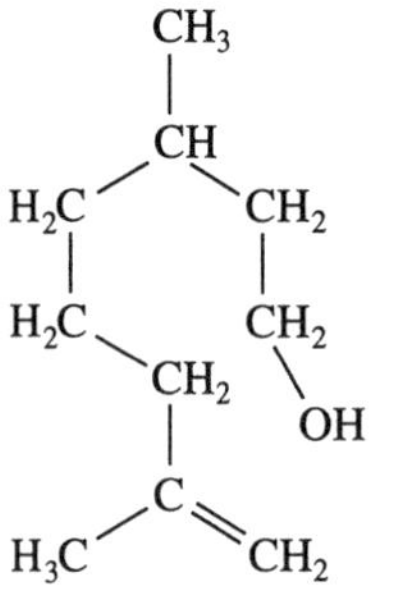

S

(a) (i) Name the functional group **common** to each of the above molecules.
(1 mark)

(ii) Describe a simple chemical test to show the presence of this functional group.
(2 marks)

(b) (i) Identify the functional group **common** to **P** and **Q** but **not** present in **R** and **S**.
(1 mark)
(ii) Describe a simple chemical test for this functional group.
(2 marks)

(c) (i) Identify the functional group **common** to **R** and **S** but **not** present in **P** and **Q**.
(1 mark)
(ii) Describe a simple chemical test for this functional group.
(2 marks)

(d) How can **Q** be reduced to **S**?
(2 marks)
[C, '92]

3. The table below shows the results of the reactions of four organic compounds **J** to **M** with the reagents shown.

Compound	Acidified potassium dichromate(VI)	2,4-Dinitro-phenylhydrazine
J	+	+
K	−	+
L	−	−
M	+	−

A + sign indicates a reaction was observed.
A − sign indicates no reaction was observed.

Simplified displayed formulae of the four compounds involved in these reactions are shown below. The formulae are numbered in random sequence.

(a) Using only the information in the table, identify the compounds **J** to **M** by giving the relevant number as shown in the diagram. In each case, show how you arrived at your conclusion by naming the functional group and commenting on the test results.
(11 marks)

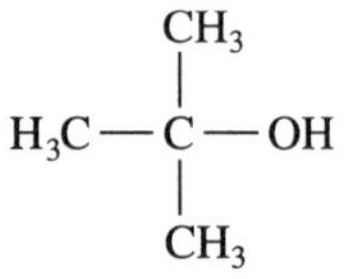

(1)

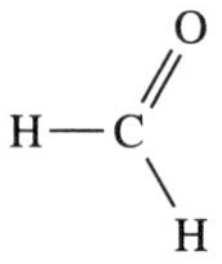

(2)

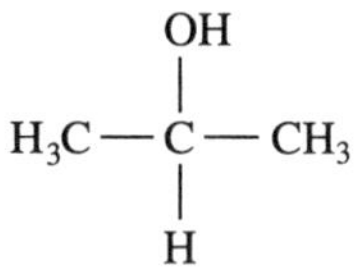

(3)

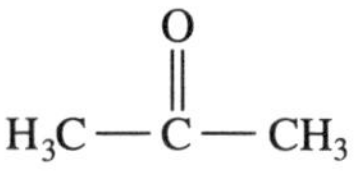

(4)

(b) (i) Draw the displayed formulae of the three structural isomers of compound (1) containing the same functional group.
(ii) Name each isomer, writing your answer beneath the relevant formula.
(6 marks)
[C, '92]

4. (a) (i) Benzene can be nitrated using a mixture of concentrated nitric acid and concentrated sulphuric acid. Suggest how the nitration might be carried out. In your account, include reference to the apparatus required and the conditions for the reaction to occur.
(10 marks)
(ii) Describe the mechanism of the nitration, including reference to the generation of the nitrating species.
(5 marks)

(b) State and explain in what ways the substitution and addition reactions of benzene with chlorine
(i) are similar to,
(ii) differ from,
those of cyclohexene with chlorine.
(10 marks)
[C, '92]

5. Benzene undergoes a number of reactions as indicated in the reaction scheme shown below.

(a) Which **two** of the numbered reactions above are examples of electrophilic substitution?

(2 marks)

(b) State the different conditions required for the **two** reactions of benzene with chlorine (reactions 2 and 3).

(2 marks)

(c) Reaction 4 is difficult to accomplish but can be achieved by using sodium hydroxide solution at 200 °C and high pressure, followed by acidification. Explain why these conditions are necessary.

(2 marks)

(d) Phenol reacts rapidly with bromine, whereas benzene reacts slowly even in the presence of a catalyst.
(i) Explain this difference in behaviour.

(2 marks)

(ii) Suggest an organic product of the reaction of phenol with bromine.

(1 mark)
[C, '95]

6. A compound **D**, C_7H_6O, reacts with 2,4-dinitrophenylhydrazine to form an orange-yellow precipitate. **D** is oxidised by acidified aqueous potassium dichromate(VI) to form the compound **E**. The reducing agent lithium aluminium hydride converts **D** to compound **F**. On treatment with aqueous sodium hydroxide, **E** gives compound **G**.

(a) Deduce the identity of compounds **D**, **E**, **F** and **G**.

(4 marks)

(b) Explain each of the reactions indicated below.
(i) **D** + 2,4-dinitrophenylhydrazine
(ii) **D** + acidified aqueous potassium dichromate(VI)
(iii) **D** + lithium aluminium hydride (lithium tetrahydridoaluminate(III))

(3 marks)

(c) What type of reaction would **D** undergo with hydrogen cyanide?

(1 mark)

(d) (i) Draw the displayed formula of the product of the reaction in (c).

(1 mark)

(ii) What form of isomerism does this compound show?

(1 mark)
[C, '95]

7. (a) Ethyl ethanoate is an example of an ester.
(i) State **three** important commercial uses of esters.
(ii) Fats and vegetable oils contain naturally-occurring esters. Give the general formula of these esters and explain why they are important commercially.

(7 marks)

(b) Describe, giving outline practical details, reaction conditions and equations, how you could produce a sample of ethyl ethanoate using bromoethane as the **only** organic starting material.

(12 marks)

(c) Esters may be hydrolysed by warming them with dilute aqueous base. The general equation for this reaction is

$$RCO_2R' + OH^- \rightarrow RCO_2^- + R'OH$$

There are **two** possible bonds which could be broken in the ester as shown in the diagram below.

Suggest how you might identify which bond, (X) or (Y), is broken in the reaction.

(6 marks)
[C, '95]

8. (a) A compound **H**, containing carbon, hydrogen and oxygen only, has the following composition by mass: C, 64.9%; H, 13.5%; O, 21.6%. **H** has two isomers **J** and **K**, which contain the same functional group. **H** does not react with acidified aqueous potassium dichromate(VI) solution, but both **J** and **K** are oxidized, **K** forming an acid. Identify **H** and its isomers **J** and **K**, explaining how you arrive at your structures.

(8 marks)

(b) Isomer **J** also consists of two isomers and these react with concentrated sulphuric acid to form a hydrocarbon, which itself exists as two isomers, **L** and **M**.
(i) Identify the type of isomerism shown by **J**, and draw the structure of its two isomers.
(ii) Write an equation for the reaction of **J** with sulphuric acid, draw the structures of the isomers **L** and **M**, and identify the form of isomerism present in these two compounds.

(8 marks)

(c) Although ethanol and phenol both contain the −OH group, they sometimes react in different ways. Give **three** ways in which these two compounds differ chemically, giving equations where appropriate.

(9 marks)
[C, '95]

4822: TRENDS AND PATTERNS

1. STATES OF MATTER

The gaseous state   Chapter 7
The basic assumptions of the **kinetic theory of gases** are that the volume of the molecules is small compared with the volume of the gas and that forces between molecules are negligible.
                          7.1, 7.2, 7.3.1

↓

The **general gas equation** or ideal gas equation:
$pV = nRT$ **7.2.6, 7.3**
can be used to determine M_r.   **1.5, 7.2.7**

↓

A real gas approaches **ideal behaviour** under certain conditions.   **7.3**
At high pressures and low temperatures, **real gases** depart from ideal behaviour.
                                       7.4

The liquid state	**Chapter 8**
Melting	8.10
Vaporisation	8.1

The solid state   **Chapter 6**
Crystalline solids may be
- simple molecular, e.g. I_2 **6.4**
- giant molecular, e.g.
  graphite,   **6.6**
  diamond,   **6.5**
  silicon(IV) oxide,   **6.7**
- metallic, e.g.   **6.2**
  aluminium and alloys,   **19.2.1**
  copper and alloys,   **24.16.1**
- hydrogen-bonded, e.g. ice.   **4.7.3**

The **properties** of a substance depend on the type of structure and bonding.

Materials are a finite resource. Recycling used materials will prolong the life of this resource.
            Additional Topic Chains and Rings 3

TRENDS AND PATTERNS

2. ELECTROCHEMISTRY

Oxidation involves an increase in **oxidation number**; reduction involves a decrease in oxidation number.
The **equation for a redox reaction** is obtained by adding **half-equations** for the changes in the oxidising agent and the reducing agent. **3.15–17**

Electrolysis is the passage of a current through a molten or aqueous compound. Substances are liberated at the electrodes. The **amounts of substances liberated** depend on the $E^\ominus$ value of the substance, the state of the electrolyte (molten or aqueous), the concentration of an aqueous electrolyte **12.2, 12.3** and the quantity of electric charge passed. **12.1.2**
The masses of solids and the masses and volumes of gases liberated can be calculated. **12.1.2**

The **Faraday constant** F is the charge on one mole of electrons = 96 500 C mol^{-1} **12.1.2**
$$F = Le$$
where L = Avogadro constant and e = charge on one electron **12.1.2**
The **Avogadro constant** can be determined by an electrolytic method.
Additional Topic Trends and Patterns 3

Listing oxidants and reductants in order of $E^\ominus$ values gives a **redox series**. **12.2**

The $E^\ominus$ values of two substances predict the **direction of any redox reaction** between them. **13.1.2**

Cells
Electrodes of different $E^\ominus$ values combine to form a voltaic cell.
$E_{\text{cell}} = E_{\text{RHS electrode}} - E_{\text{LHS electrode}}$
A positive value of E_{cell} indicates that the **cell reaction** will happen spontaneously and that electrons will flow from left to right through the external circuit. **13.1**

Industrial uses of electrolysis include
- the electrolysis of brine, **12.4, 18.8.2**
- the extraction and anodising of aluminium, **12.4, 18.8.2, 19.2.1**
- the purification of copper. **12.4, 24.16**

The **rusting of iron** is an electrochemical reaction. **24.14.4, 24.14.5**

For a **redox system**,
Oxidant + ne^- ⇌ Reductant
An electrode immersed in the system acquires a potential that depends on the position of the equilibrium between the oxidant and reductant. **3.15–17**

The **standard electrode potential** $E^\ominus$ is the potential of an electrode in a 1 mol dm^{-3} solution of ions at 298 K. **13.1, 13.1.1**

An electrode potential is measured by finding the emf of a cell composed of the electrode and a standard electrode, e.g. the **standard hydrogen electrode**, which has $E^\ominus = 0$ by definition, or the calomel electrode. **13.1.2**

Chemical cells which are being developed for more widespread use include fuel cells and improved batteries for electric vehicles.
13.3.4 and Additional Topic Trends and Patterns 2

35

3. GROUP TRENDS IN THE PERIODIC TABLE

3.1 THE GROUP 2 ELEMENTS AND THEIR COMPOUNDS

Group 2 is the **alkaline earth metals**, Be, Mg, Ca, Sr, Ba. They have strongly basic oxides. The chlorides are crystalline solids. They form salts including sulphates, nitrates, carbonates and ionic hydrides. **18.1**

Trends down Group 2
1. **First ionisation energy** decreases down the group. **18.1**
2. Passing down Group 2,
   - solubility of MSO_4 decreases as $\Delta H^{\ominus}$ of hydration of M^{2+} decreases with increase in size of M^{2+}. **18.5.2**
   - solubility of $M(OH)_2$ increases down the group as lattice enthalpy decreases with increase in size of M^{2+}. **18.5.2**
3. **Nitrates** $M(NO_3)_2$ decompose on heating to give $MO + NO_2 + O_2$. **18.5.3**
4. **Carbonates** MCO_3 decompose on heating to give $MO + CO_2$. **18.5.3**
5. **Hydrogencarbonates** $M(HCO_3)_2$ exist only in solution. **18.5.3**
6. **Hydroxides** $M(OH)_2$ form oxides $MO + H_2O$ on heating. **18.5.3**

Note: Chapter 18 deals with Group 1 and Group 2. **UCLES** candidates are required to know about sodium and Group 2.

Properties
Group 2 are harder and denser than Group 1 with higher melting and boiling temperatures. They are softer than transition metals with lower melting temperatures. They form ionic compounds in which the metals have oxidation number $= +2$. **18.1**

Reactions
From the first member to the last, there is an increase in reactivity down the group.

Reaction with oxygen
All react with oxygen, tarnish in air and burn readily to form oxides. **18.4.2, 18.4.3**

Reaction with halogens
All react with halogens on heating to form halides. **18.4.3**

Reaction with water
All, except Be and Mg, react with cold water to form MOH or $M(OH)_2 + H_2$. **18.4.2**

Reaction of oxides with water
The oxides react with water to form solutions of hydroxides. Group 2 hydroxides are less soluble than Group 1 hydroxides, e.g. NaOH. **18.5.2, 18.5.4** omitting peroxides and superoxides

Important uses of Group 2 compounds include
- MgO as a refractory lining in furnaces **18.5.5 and Additional Topic Trends and Patterns 3**
- CaO and $Ca(OH)_2$ in agriculture to neutralise excessive acidity. **18.5.5**

3.2 GROUP 4

GROUP 4: C, Si, Ge, Sn, Pb

Electrical conductivity
Graphite is an electrical conductor. Silicon and germanium are semiconductors. Tin and lead are conductors. **23.1, 23.2, 23.4**

Melting temperatures
Carbon and silicon have high melting temperatures because they consist of macromolecular structures. Tin and lead have lower melting temperatures than most metals. **23.2**

Oxidation states in oxides and aqueous cations
C $+2$ and $+4$ (in e.g. CO and CO_2)
Si $+4$ (in e.g. SiO_2)
Ge $+4$ (in e.g. GeO_4)
Sn $+2$ and $+4$ (Sn(II) is a reducing agent; Sn(II) compounds are ionic; Sn(IV) compounds are covalent)
Pb $+2$ and $+4$ (Pb(IV) is an oxidising agent; Pb(II) compounds are ionic; Pb(IV) compounds are covalent). **23.1, 23.6.3, 23.6.4, 23.8**

Bonds formed in Group 4
Metallic character increases down the group. There is an increasing tendency down the group for a pair of valence electrons to be 'inert' and for ionic bonds to be formed by E^{2+}.
The 'inert pair effect' increases down the group. Sn(II) compounds are reducing agents because Sn(IV) is more stable than Sn(II), but Pb(IV) compounds are oxidising agents because Pb(II) is more stable than Pb(IV).
Carbon forms no **complex ions**. Later elements in the group can accept electrons from ligands in their M shells to form complex ions, e.g. $PbCl_6^{2-}$. **23.1, 23.2**
Carbon alone forms π bonds. **23.6.2**

Ceramics based on silicon(IV) oxide are important materials with many uses.
Additional Topic Trends and Patterns 3

Oxides
Bonding is covalent in the oxides EO_2. The oxides SnO and PbO are largely ionic in character. **26.1–3**
Thermal stability
SnO_2 is more stable than SnO
$PbO_2 \xrightarrow{\text{heat } 400°C} Pb_3O_4$ **23.7.4**
Acid/base nature
CO neutral, but reacts with molten bases
CO_2 acidic
SiO_2 reacts with molten bases
SnO, SnO_2, PbO, PbO_2 amphoteric (SnO and PbO are more basic than SnO_2 and PbO_2) **23.7.3**

Carbon and carbon monoxide are important **reducing agents**. **23.7.3**
Carbon monoxide reduces iron oxides in the blast furnace. **24.14.1**

Halides
(**E** = element in Group 4, X = halogen)
In Group 4, EX_4 are all covalent and, except for CX_4, are readily hydrolysed. CX_4 is not hydrolysed because C has a full L shell and has no M shell into which H_2O can coordinate as a first step in hydrolysis.
SnX_2 and PbX_2 are ionic. **23.6.4, 23.7.2**

3.3 GROUP 7

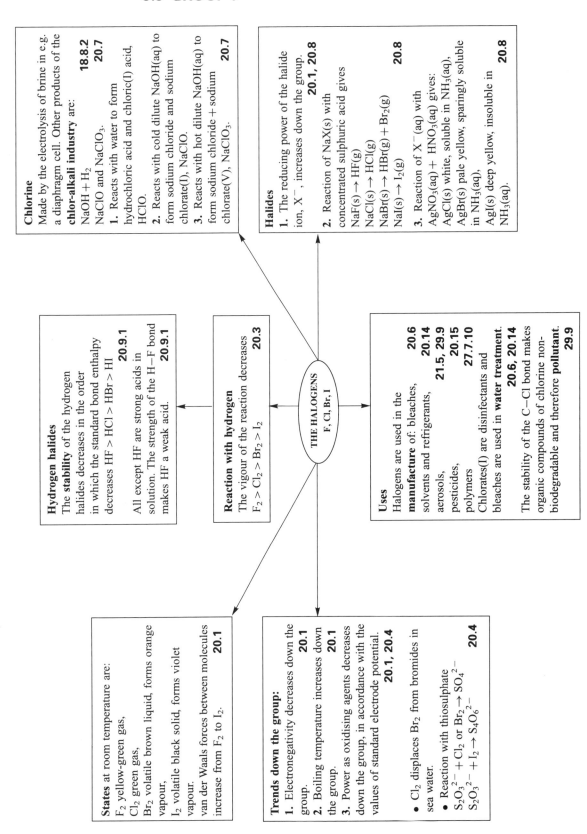

Chlorine
Made by the electrolysis of brine in e.g. a diaphragm cell. Other products of the **chlor-alkali industry** are:
$NaOH + H_2$ **18.8.2**
NaClO and NaClO$_3$. **20.7**

1. Reacts with water to form hydrochloric acid and chloric(I) acid, HClO.
2. Reacts with cold dilute NaOH(aq) to form sodium chloride and sodium chlorate(I), NaClO.
3. Reacts with hot dilute NaOH(aq) to form sodium chloride + sodium chlorate(V), NaClO$_3$. **20.7**

Halides
1. The reducing power of the halide ion, X$^-$, increases down the group. **20.1, 20.8**
2. Reaction of NaX(s) with concentrated sulphuric acid gives
NaF(s) → HF(g)
NaCl(s) → HCl(g)
NaBr(s) → HBr(g) + Br$_2$(g)
NaI(s) → I$_2$(g) **20.8**
3. Reaction of X$^-$ (aq) with AgNO$_3$(aq) + HNO$_3$(aq) gives:
AgCl(s) white, soluble in NH$_3$(aq),
AgBr(s) pale yellow, sparingly soluble in NH$_3$(aq),
AgI(s) deep yellow, insoluble in NH$_3$(aq). **20.8**

Hydrogen halides
The **stability** of the hydrogen halides decreases in the order in which the standard bond enthalpy decreases HF > HCl > HBr > HI **20.9.1**
All except HF are strong acids in solution. The strength of the H–F bond makes HF a weak acid. **20.9.1**

Reaction with hydrogen
The vigour of the reaction decreases
F$_2$ > Cl$_2$ > Br$_2$ > I$_2$ **20.3**

THE HALOGENS
F, Cl, Br, I

Uses
Halogens are used in the **manufacture** of: bleaches, **20.6**
solvents and refrigerants, **20.14**
aerosols, **21.5, 29.9**
pesticides, **20.15**
polymers **27.7.10**
Chlorates(I) are disinfectants and bleaches are used in **water treatment**. **20.6, 20.14**
The stability of the C–Cl bond makes organic compounds of chlorine non-biodegradable and therefore **pollutant**. **29.9**

States at room temperature are:
F$_2$ yellow-green gas,
Cl$_2$ green gas,
Br$_2$ volatile brown liquid, forms orange vapour,
I$_2$ volatile black solid, forms violet vapour.
van der Waals forces between molecules increase from F$_2$ to I$_2$. **20.1**

Trends down the group:
1. Electronegativity decreases down the group. **20.1**
2. Boiling temperature increases down the group. **20.1**
3. Power as oxidising agents decreases down the group, in accordance with the values of standard electrode potential. **20.1, 20.4**

- Cl$_2$ displaces Br$_2$ from bromides in sea water.
- Reaction with thiosulphate
$S_2O_3^{2-} + Cl_2$ or $Br_2 \rightarrow SO_4^{2-}$
$S_2O_3^{2-} + I_2 \rightarrow S_4O_6^{2-}$ **20.4**

MODULE 4822: TRENDS AND PATTERNS

1. States of matter

(a) the kinetic theory as applied to an ideal gas	7.1–3	(f) describe and interpret the uses of	
		(i) aluminium and its alloy duralumin	19.2.1
(b) departure from ideal behaviour at high pressures and low temperatures	7.3, 7.4	(ii) copper and its alloy brass	24.16.1
(c) the general gas equation $pV = nRT$, including calculations	7.2.6, 7.2.7, 7.3	(g) materials as a finite resource; the importance of recycling	See **Additional Topic Chains and Rings 3**
(d) a kinetic-molecular model of the liquid state, melting and vaporisation	8.1, 8.2		
(e) explain the high melting point, electrical insulating and thermal insulating properties of silicon(IV) oxide	27.7.3	(h) suggest from physical data the type of bonding in a substance	4.7.3

2. Electrochemistry

(a) redox processes	3.15–17	(i) calculations on electrolysis	**12.1.2** and **Checkpoint 12A**
standard electrode potential and standard cell potential	13.1, 13.1.1	(j) (i) the development of cells, e.g. H_2/O_2 fuel cell, improved batteries for electric vehicles	See **Additional Topic Trends and Patterns 2**
(b) calculate a standard cell potential from two standard electrode potentials	13.1.2		
(c) establish a reactivity series of metals from their reactions; relate the series to their standard electrode potentials	12.3.1	(ii) lead–acid cells, dry cells, mercury and lithium batteries	13.3
(d) use standard electrode potentials to explain/predict the direction of electron flow in a cell and the feasibility of a reaction	13.1.2	(k) (i) the electrolysis of brine, using a diaphragm cell	12.4, 18.8.2
(e) construct redox equations	3.15	(ii) the extraction of aluminium from molten aluminium oxide	12.4, 19.2.2
(f) $F = Le$ (F = Faraday constant, L = Avogadro constant, e = charge on the electron)	12.1.2	(iii) the electrolytic purification of copper	12.4, 24.16
(g) determination of the Avogadro constant from an electrolytic measurement	See **Additional Topic Trends and Patterns 1**	(l) the electrochemical nature of the rusting of iron	24.14.4
		(m) methods of protecting iron from rusting	24.14.5
(h) factors that affect the identity of the substance liberated during electrolysis	12.2, 12.3		

3. Group trends in the Periodic Table

3.1 The Group 2 elements and their compounds

Note that Chapter 18 of A-Level Chemistry covers Groups 1 and 2. UCLES students are required to study Group 2 and to study some aspects of sodium as the Group 1 member of Period 3.

(a) the reactions of the elements with oxygen and water	18.4.2, 18.4.3	(c)(d) the thermal decomposition of the nitrates and carbonates; explanation	18.5.3
		(e) the variation in solubility of the sulphates; explanation	18.5.2
(b) the behaviour of the oxides with water	18.5.2, 18.5.4 (omitting peroxides and super-oxides)	(f) the trends in properties of the elements and their compounds	18.1, 18.5.1
		(g) the use of magnesium oxide as a refractory lining and of calcium carbonate as a building material	18.5.5 and see **Additional Topic Trends and Patterns 3**
		(h) the use of lime in agriculture	18.5.5

3.2 The Group 4 elements and their compounds		(*d*)(*e*) reactions with hydrogen; stabilities of the hydrides	**20.3** including **Table 20.3** and **20.9.1**
(*a*) the variation in melting point and in electrical conductivity of the elements related to structure and bonding	**23.2**		
(*b*)(*c*) the tetrachlorides: bonding, molecular shape, volatility, reactions with water	**23.6.4, 23.7.2**	(*f*) the reactions of halide ions with (i) aqueous silver ions followed by aqueous ammonia, (ii) concentrated sulphuric acid	**20.8, Table 20.5**
(*d*) the oxides of oxidation states II and IV: bonding, acid–base nature and thermal stability	**23.6.1, 23.6.3, 23.7.3**	(*g*) the reaction of chlorine with cold, and with hot, aqueous sodium hydroxide	**20.7** (Note that only chlorine is mentioned in the syllabus.)
(*e*) the relative stability of different oxidation states of the elements in their oxides and aqueous cations, with reference to $E^\ominus$ values.	**23.1, 23.6.3, 23.6.4, 23.8**	(*h*) the use of chlorine in water purification	**20.6, 20.14**
3.3 The Group 7 elements and their compounds		(*i*) the industrial importance and environmental significance of the halogens and their compounds	**20.6, 27.7.10, 20.14, 21.5, 29.9**
(*a*)(*b*) the trends in volatility and colour of chlorine, bromine and iodine	**20.1**		
(*c*) the elements as oxidising agents, with reference to $E^\ominus$ values.	**20.4** including **Checkpoint 20B**		

ADDITIONAL TOPIC TRENDS AND PATTERNS 1: DETERMINATION OF THE AVOGADRO CONSTANT BY AN ELECTROLYTIC METHOD

One mole of any substance contains the Avogadro constant of particles per mole (see **3.5**). One mole of ions, electrons or atoms contains the Avogadro constant of particles.

The charge on a singly charged ion, e.g. Ag^+, is the same as the charge on an electron (**12.1.1**).

$$Ag^+(aq) + e^- \rightarrow Ag(s)$$

• One singly charged ion, e.g. Ag^+, requires one electron to discharge it as an atom.

• One mole of singly charged ions, e.g. Ag^+, therefore requires one mole of electrons for discharge.

• The quantity of electricity required to discharge one mole of singly charged ions in electrolysis can be found by experiment.

• This quantity of electricity is therefore the quantity of electric charge per mole of electrons. It is called the Faraday constant. You can refer to **12.1.1** for experimental details of the determination of the Faraday constant. The value is 9.649×10^4 C mol^{-1} (C = coulomb).

The basis of the method is ...

... Faraday constant F = electric charge per mole of substance ...

... Avogadro constant L = number of particles per mole of substance ...

... e = charge on one electron ...

Therefore F = Le

Since

Faraday constant = electric charge per mole of substance,

and

Avogadro constant = number of particles per mole of substance

Faraday constant/Avogadro constant = electric charge/number of particles

= electric charge on one particle

= charge on one electron

That is,

$$F = Le$$

where F = Faraday constant, L = Avogadro constant, e = electronic charge

Since e is known ...
... and F can be found by experiment ...
... L can be calculated.

The charge on one electron is known. It was determined by Millikan in his oil drop experiment. This is described in physics texts, e.g. *A-Level Physics* by Roger Muncaster, Fourth edition (Stanley Thornes), **50.4**. The value is 1.602×10^{-19} C. Substituting values of F and e into the equation

$$L = F/e$$
$$L = 9.649 \times 10^4 \text{ C mol}^{-1} / 1.602 \times 10^{-19} \text{ C} = 6.022 \times 10^{23} \text{ mol}^{-1}.$$

The value of the Avogadro constant is 6.022×10^{23} mol^{-1}.

ADDITIONAL TOPIC TRENDS AND PATTERNS 2: IMPROVED BATTERIES

(Read after **13.3.4**)

1. CHARACTERISTICS OF BATTERIES

The most important characteristics of a battery are:

Batteries store electrical energy by means of a reversible electrochemical reaction.

- storage capacity, the quantity of charge available without recharging
- energy density, the electrical energy available from a battery in one cycle of operation
- power density, the electrical power available from a battery in one cycle of operation

Storage capacity decreases with the age of the battery. This is a result of changes in the structure of the electrodes; for example in the lead–acid battery lead sulphate may block part of the lead electrode.

The lead-acid battery (13.3.3) has a typical energy density of 15 Wh kg^{-1} and a power density of about 150 W kg^{-1}, and a voltage of 2.04 V. If batteries are to play a major role in transportation, these characteristics must be improved. If battery-operated vehicles are to become suitable for widespread use, they will require smaller, lighter batteries of higher voltage.

They include ...
... the lead-acid battery ...
... the nickel-cadmium battery ...
... the silver-zinc battery ...
... the sodium-sulphur battery.

Nickel-cadmium batteries have voltage 1.5 V, energy density 40 Wh kg^{-1} and power density of 150 W kg^{-1}. They are an improvement on the lead–acid battery, but the price of nickel and cadmium makes these batteries more expensive than lead–acid batteries.

Silver-zinc batteries have high energy densities of 110 Wh kg^{-1}, and are already suitable for use in electric cars, but the price of silver makes them expensive.

Sodium-sulphur batteries have been developed by the Ford Motor Company. With energy density 200 Wh kg^{-1}, they can store over ten times more energy than lead–acid batteries of the same mass. The electrolyte is a piece of solid alumina. Since the anode is liquid sodium and the cathode is liquid sulphur, the cell must operate at 250–400 °C.

Many other redox systems have been utilised in the construction of batteries. The incentive to replace the petrol-driven vehicle is strong because of pollution from the internal combustion engine (**26.4**) and because the Earth's reserves of crude oil are limited. Research continues into battery technology.

2. FUEL CELLS

In a battery the materials that take part in the electrochemical reaction are all contained in the unit. The cell reaction can be reversed at will to recharge the battery or to discharge the battery. In a fuel cell, on the other hand, the reactants are supplied from outside the cell and often the reaction is difficult to reverse. A fuel cell converts chemical energy into electrical energy by electrochemical reactions in a galvanic cell with a continuous supply of reactants. The hydrogen–oxygen fuel cell is described in **13.3.4**, including Figure 13.9. It has a voltage of 1.2 V, an energy density of 1000–2000 Wh kg^{-1} and power density of 100 W kg^{-1}

Performance of Some Batteries and Fuel Cells

Cell	Potential/V	Typical energy density/Wh kg^{-1}	Typical power density/W kg^{-1}
Lead–acid cell	2.04	15	150
Nickel–cadmium cell	1.48	40	150
Silver–zinc cell	1.70	110	
Sodium–sulphur cell		200	
Hydrogen–oxygen fuel cell	1.20	1000–2000	100

Fuel cells use a continuous supply of reactants in an electrochemical reaction which generates electrical energy.

Any ionic redox reaction can in theory be used in a fuel cell. The ease of handling of the fuel must be considered, and its reaction products should be manageable. Other possibilities are the oxidation by oxygen of hydrazine (N$_2$H$_2$), carbon, carbon monoxide, methanol, ammonia and methane. In practice the reaction must take place at a reasonable rate and the change in free energy must be high so that the voltage developed is adequate. In **13.1.2**, you met the equation

$$\Delta G = -nzFE^{\ominus}$$

where ΔG = change in free energy, n = number of moles which react, z = number of electrons transferred per equation of the reaction, F = Faraday constant, $E^{\ominus}$ = standard electrode potential. For a cell, the energy output is ΔG.

$$\Delta G = -nzFE$$

where E = cell voltage.

They are very efficient in energy conversion.

The main attraction of fuel cells is the high efficiency of energy conversion. The energy output of a fuel cell ΔG can be compared with the maximum amount of energy that can theoretically be obtained from the fuel, that is the enthalpy of combustion of the fuel, ΔH for the cell reaction. For the hydrogen–oxygen fuel cell $\Delta H = -242$ kJ mol^{-1}, while $\Delta G = -229$ kJ mol^{-1}. Thus the thermal efficiency of the cell is 229/242 or 95%. The overall efficiency of the cell is lower than this because of imperfections in the electrodes, the resistance of the electrolyte and the occurrence of side reactions. In practice an efficiency of about 80% is achieved in fuel cells.

3. ELECTROCHEMICAL TRANSPORT

Batteries are in use as power supplies in transport vehicles … … but the lead-acid battery needs recharging after a short distance …

One of the most important applications of batteries and fuel cells is as power units for vehicles. Already thousands of fork-lift trucks and milk delivery vans are powered by lead-acid storage batteries. The range of these vehicles is limited by the low energy density of the lead–acid battery, and their uses are limited. For a milk delivery van with a short delivery round, a battery is better than a petrol engine because the electric motor can be switched on and off with no trouble whereas a petrol engine must be left to idle. The main problem is the speed of recharging batteries, and new battery chargers can recharge the lead–acid battery in 3 hours instead of the usual 8 hours.

A small saloon car weighing 1 tonne requires 20 kWh of energy to travel 55 km (35 miles). It can do this on 4 kg (1 gallon) of petrol. An electrically operated car would require 1.5 tonne of lead–acid batteries, occupying 0.5 m^3 of space. The prototype electric cars are two-seaters with a range of 65 km (40 miles) and maximum speed 100 km h^{-1}.

... and widespread use of electric vehicles awaits improvements in battery technology.

Storage batteries have good power densities (W kg^{-1}) but poor energy densities (Wh kg^{-1}). Fuel cells have poor power densities but good energy densities. A power source consisting of a combination of batteries and fuel cells may be the answer. During steady cruising the fuel cells would supply power to the electric motor and also recharge the batteries. On steep gradients and at times when rapid acceleration is needed, as in starting up and for overtaking, the batteries could be used to provide bursts of power.

CHECKPOINT ON ADDITIONAL TOPIC TRENDS AND PATTERNS 2

1. Discuss the suitability of lead–acid batteries for town bus services.

2. What is the disadvantage of lead–acid batteries for small cars?

3. List some characteristics of fuel cells.

ADDITIONAL TOPIC TRENDS AND PATTERNS 3: CERAMICS

WHAT ARE THEY?

The refractory bricks with which furnaces are lined are ceramics.

The blast furnace in which iron ore is smelted and the kiln in which cement is made are lined with refractory (heat-resisting) bricks. These refractory bricks contain substances such as silicon oxide or magnesium oxide or aluminium oxide. These refractory materials are **ceramics** and have been known and used for thousands of years. Traditional ceramics such as these are derived from the raw materials clay and silica and are essentially **silicates**.

Modern ceramics can function at higher temperatures than metals.

They are hard and wear-resistant.

Ceramics are crystalline compounds of oxygen, carbon or nitrogen with other elements, e.g. silicon, aluminium and some transition metals.

In recent years new ceramics have been developed to meet the demand for materials which withstand high temperatures and high pressures and resist chemical corrosion. **Ceramics comprise all inorganic engineering materials and products, except for metals and alloys.** Many ceramics have good mechanical strength, are hard and resistant to wear. Their low conductivity of heat and electricity makes them useful as building bricks, in electrical insulation and for a variety of other applications. Modern ceramics, e.g. silicon nitride and silicon carbide, can function at higher temperatures than alloy steels and are used as components of the internal combustion engine. The inclusion of cermet (ceramic and metal) parts and ceramic parts and insulating ceramic coatings can raise the operating temperature of a diesel engine from 700 °C to 1100 °C, which increases the engine's efficiency by 50%. These ceramics are made by compressing hot mixtures of finely powdered oxides. Alternatively the powders may be sprayed on to a metal to give a hard coating. Examples of modern ceramics are soda glass, silica glass, aluminium oxide, zirconium oxide, titanium carbide, zirconium carbide, tungsten carbide, silicon carbide and silicon nitride.

STRENGTH

They are brittle, but ceramic fibres are very strong.

Ceramics are brittle, with low tensile strength and low toughness. They can fail at low stress. Because the toughness is so low, small defects, e.g. surface scratches or lumps, can reduce the strength drastically. On the other hand, ceramic fibres, e.g. glass fibres, aluminium oxide fibres, silicon carbide fibres, are among the strongest materials known. This is because their surface defects are very small – a few μm only. The search is on for fibres which will reinforce ceramics. The hope is to achieve the same sort of success as fibre-reinforced plastics.

USE OF CERAMICS

Whiteware is a term for ceramic products which are usually white and of fine texture. Whiteware includes earthenware, chinaware, porcelain, sanitary ware, stoneware and floor tiles.

Ceramics include whiteware, structural clay products, refractories, heat-resistant ceramic tiles for spacecraft, engineering components and tools.

Structural clay products include building brick, face brick, terracotta, sewer pipe and drain tile. They are manufactured from the cheapest of common clays.

Refractory ceramics can withstand the thermal, chemical and physical effects encountered in furnaces. They can be used as furnace linings at much higher temperatures than metals. Some have a very high silicon(IV) oxide content, and others have a very high aluminium oxide content.

Ceramics in space When a space shuttle re-enters the Earth's atmosphere from space, the temperature of the surface can reach 1500 °C – hotter than molten steel. A covering layer of heat-resistant ceramic tiles insulates the interior of the space shuttle so as to protect the computers and recording instruments inside.

Engineering components A ceramic coating on cast iron improves resistance to wear and tear and heat and therefore extends the life of engine parts. Research is going on into the use of ceramics to allow diesel engines and jet engines to run at higher temperatures. The higher the temperature at which an engine operates, the more efficiently it runs and the less fuel it consumes.

Gas turbine blades made of ceramics, e.g. silicon nitride, can run at higher temperatures than alloys.

Machine tools made of ceramics can rotate twice as fast as metal tools without deforming or wearing out.

STRUCTURE

The bonding in ceramics may be ionic or covalent. The bonds are directed in space, and the structure is therefore more rigid than a metallic structure.

Ceramics consist of a regular arrangement of atoms. The repeating structural units are atoms or ions covalently bonded molecules arranged in a regular three-dimensional structure. The bonding may be covalent, e.g. diamond and silicon(IV) oxide, ionic, e.g. magnesium oxide, or intermediate between pure ionic and pure covalent, e.g. aluminium oxide. Ceramics include glasses, which have a solid structure in which regular crystallinity has been disordered.

*This structure makes ceramics …
… crystalline …
… hard … brittle …
… of high melting temperature …
… poor conductors of heat and electricity because they lack free electrons …
… usually opaque because light is reflected by grain boundaries in the crystal structure.*

The bonds between atoms in a ceramic are more rigid than in metals so that plastic deformation takes place with difficulty. Ceramics are therefore usually rather brittle. Metals and ceramics represent two extremes of crystalline behaviour, from metals with good ductility and high toughness to ceramics with poor ductility and high toughness. Most ceramics are much harder than metals and have higher melting temperatures. This is due to the powerful directional bonding in these materials. Ceramics are poor conductors of heat and electricity because they lack free electrons.

Silicon(IV) oxide, silica, SiO_2, is the basis of a large variety of ceramics. A silicon atom can form covalent bonds with four oxygen atoms to form the unit SiO_4^{4-} with a tetrahedral distribution of bonds (see Figure 1).

Each of the four oxygen atoms in SiO_4^{4-} has an unshared electron which it can use to combine with other SiO_4^{4-} units. Quartz and other crystalline forms of silicon(IV) oxide have a three-dimensional network of SiO_4^{4-} units, in which all four oxygen atoms of the silicate tetrahedra are share with other Si atoms (see Figure 2). With each silicon atom bonded to four oxygen atoms and each oxygen atom bonded to two silicon atoms, the formula is $(SiO_2)_n$.

FIGURE 1 The SiO_4^{4-} Tetrahedral Ion

● = silicon   ○ = oxygen

The outline of the tetrahedron is shown.
The bonds between the Si atom and the O atoms are not shown.

FIGURE 2 The Three-dimensional Network of Silicate Tetrahedra in Quartz, SiO_2

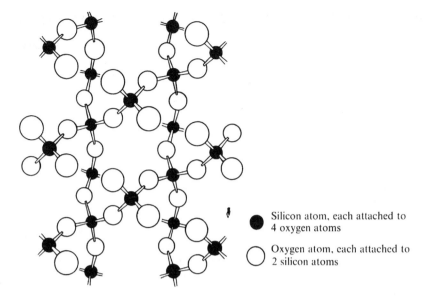

- Silicon atom, each attached to 4 oxygen atoms
- Oxygen atom, each attached to 2 silicon atoms

In quartz, all the tetrahedral structures have silicon atoms, but in other minerals up to half the silicon atoms can be replaced by aluminium atoms. For every aluminium atom that replaces a silicon atom, an ion, e.g. Na^+ or K^+, is needed to balance the charge, as in feldspar, $KAlSi_3O_8$. The chief component of the Earth's crust is granite, which is a mixture of mica, feldspar and quartz.

GLASSES

Glasses are amorphous, non-crystalline solids composed of linked silicate chains.

When molten silica is cooled very slowly, it crystallises at the freezing temperature. If molten silica is cooled more rapidly, it is unable to organise all the atoms into the ordered arrangement required for a crystal and it solidifies as a disordered arrangement which is called a **glass**. Glasses are amorphous, disordered non-crystalline substances composed of linked silicate chains (see Figure 3).

Glass is transparent. The reason is that glass is amorphous: the arrangement of atoms is disordered and provides no reflecting surfaces, such as the grain boundaries which make metals opaque. Crystalline ceramics, on the other hand, have regularly spaced layers of atoms which reflect light, and these ceramics are opaque.

FIGURE 3 A Glass – a Disordered Chain of Silicate Tetrahedra

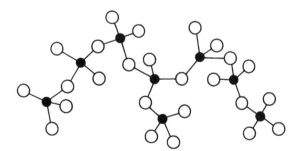

CHECKPOINT ON ADDITIONAL TOPIC TRENDS AND PATTERNS 3

1. (*a*) Show the electron structure of SiO_4^{4-} by means of a 'dot and cross' diagram.

(*b*) Show how this unit can act as the building block of silicate minerals.

2. (*a*) Why are glasses classified as ceramics?

(*b*) How do glasses differ from quartz in properties?

(*c*) How are these differences related to the difference in structure?

(*d*) Why is glass transparent and china opaque?

3. Explain why ceramics are (*a*) harder than metals, (*b*) more brittle than metals, (*c*) generally poor electrical conductors.

EXAMINATION QUESTIONS ON TRENDS AND PATTERNS

1. (*a*) Give an account of
(i) the formulae and bonding,
(ii) the physical properties,
of the oxides of the Period Three elements (Na to Ar inclusive).
(12 marks)

(*b*) How do these oxides react with water? Give equations where relevant.
(13 marks)
[C, '95]

2. (*a*) The halogens chlorine, bromine and iodine show a gradation of properties in their volatility and in their reaction with sodium thiosulphate. Describe and explain these gradations.
(7 marks)

(*b*) Describe and interpret the behaviour of the hydrides of chlorine, bromine and iodine when they are added to water.
(6 marks)

(*c*) Refer to tables for relevant $E^\ominus$ values and use these to explain the displacement reactions of the halogens.
(5 marks)

(*d*) Astatine is a member of the halogen family.
(i) Predict **three** chemical properties of astatine and explain your reasoning by reference to the properties of the other halogens.
(ii) Suggest why astatine is not normally studied in school or college laboratories.
(7 marks)
[C, '95]

(**Table 13.1** *gives $E^\ominus$ values.*)

3. (*a*) State the relationship between the Faraday constant, F, the Avogadro constant, L, and the charge on an electron, e.
(1 mark)

(*b*) A solution of sulphuric acid was electrolysed for 2.00 hours, during which time a constant current of 0.040 A was flowing.
(i) Name the gas produced at each electrode.
(2 marks)
(ii) What quantity of electricity (in coulombs) flowed through the acid?
(1 mark)
(iii) What amount (in moles) of electrons flowed through the acid?
(1 mark)
(iv) What amount (in moles) of gas was evolved at the cathode?
(1 mark)
(v) What volume of gas was evolved at the cathode?
(Under the experimental conditions used, the molar volume is 24 dm^3.)
(1 mark)
[C, '95]

4. (*a*) (i) Copy and complete the table below the show the formulae and structures of the oxides of the given elements, where the elements are in their highest oxidation state.

Element	Mg	Al	Si	P	S
Formula of oxide	MgO	Al$_2$O$_3$		P$_4$O$_{10}$	
Structure of oxide	Giant ionic				

(4 marks)

(ii) Describe the trend in behaviour of the oxides in the table with:
A. water;
(3 marks)
B. an acid;
(2 marks)
C. an alkali.
(3 marks)

(*b*) State and explain which of the oxides Na$_2$O, MgO or Al$_2$O$_3$ is likely to have the highest melting point.
(3 marks)
[C, '93]

5. Iron and copper are two important metals.
(*a*) Complete the electronic configurations of
(i) the iron atom;
(ii) the copper(I) ion, Cu$^+$.

Fe 1s^2 2s^2 2p^6 3s^2 3p^6
(1 mark)

Cu$^+$ 1s^2 2s^2 2p^6 3s^2 3p^6
(1 mark)

(*b*) Iron forms two stable positively charged ions, but Group I metals, e.g. sodium, form only one ion.
(i) Explain why iron can form both Fe^{2+} and Fe^{3+} ions.
(2 marks)
(ii) Explain why Group I metals form only one stable ion with a single positive charge.
(1 mark)

(*c*) What is the oxidation state of
(i) copper in Cu$_2$O?
(1 mark)
(ii) iron in K$_3$FeCl$_6$?
(1 mark)

(*d*) (i) What do you understand by the term *complex ion*?
(2 marks)
(ii) Give the formulae of **two** complex ions of copper in its oxidation state of $+2$.
(2 marks)
[C, '93]

6. Refer to the electrode potentials given in tables when answering this question.
(*a*) Indicate whether the following reactions would be expected to occur.
(i) Mg(s) + Pb^{2+}(aq) → Mg^{2+}(aq) + Pb(s)
(1 mark)
(ii) Fe(s) + Ba^{2+}(aq) → Fe^{2+}(aq) + Ba(s)
(1 mark)

(*b*) (i) By calculating its $E^\ominus$ value, show that the following reaction between hydrogen gas and copper(II) ions would be expected to occur.

$$Cu^{2+}(aq) + H_2(g) \to Cu(s) + 2H^+(aq)$$

(1 mark)
(ii) Explain why, in fact, no reaction occurs when hydrogen gas is bubbled through aqueous copper(II) sulphate.
(1 mark)

(*c*) A simple cell can be constructed using a Zn/Zn^{2+} half-cell and a Cu/Cu^{2+} half-cell.
(i) Write an equation for the process occurring at the positive electrode (anode).
(1 mark)

(ii) Write an equation for the process occurring at the negative electrode (cathode).
(1 mark)
(iii) Write an equation for the overall cell reaction.
(1 mark)
(iv) In which direction will the electrons flow in the external circuit?
(1 mark)
(v) What is the value of E for this cell at 25 °C, when $[Zn^{2+}] = [Cu^{2+}] = 1$ mol dm^{-3}? See **Table 13.1** for $E^\ominus$ values.
(1 mark)

(vi) State whether the value of E for the cell will increase, decrease or stay the same when the concentrations in each half cell are as shown below.
1. $[Zn^{2+}] = 1$ mol dm^{-3} and $[Cu^{2+}] = 2$ mol dm^{-3}
(1 mark)
2. $[Zn^{2+}] = 2$ mol dm^{-3} and $[Cu^{2+}] = 1$ mol dm^{-3}
(1 mark)
[C, '95]

4823: MATERIALS

Note that the references in this module are to *Materials Science* by E. N. Ramsden (Stanley Thornes, 1995).

1. CERAMICS

Structure
The **elements** present may be metallic or non-metallic. The structural units are ionically or covalently bonded and arranged in a **three-dimensional structure**.   **3.5, 3.5.1–4**
Ceramics may be crystalline, amorphous or bonded.   **3.1.1**
They include complex silicates   **3.5**
and cements.   **3.6.1, 3.6.2**
The tetrahedral SiO_4^{4-} unit is a **building block** for silicates.   **3.5**
The structure can be **modified**, e.g. silica changed into an amorphous **glass**.
    3.5.5, 6.2.2, 6.2.3, 6.2.7

Processing
Ceramics are made by compressing powders into shape and then firing them.   **3.2, 3.3**
Cements and **concretes** are made in bulk.   **3.6.1–3**
Glasses are produced as melts. Devitrification occurs slowly below the **glass transition temperature** or **glass transformation temperature**.
Annealing reduces stresses in glass.
    3.5.5, 3.7–10
The addition of metal compounds makes **coloured glasses**.   **3.11**

Properties
Ceramics are hard, with high melting temperatures, are electrical and thermal insulators, are opaque, rather brittle and are resistant to corrosion. They are **very stable** at high temperatures.
    3.1, 3.5, 3.6.4, 4.3
The brittleness of ceramics is caused by **Griffiths cracks**.   **1.10**

Uses
Uses of ceramics include: whitewares, bricks, tiles and pipes, firebricks for furnace linings, enamels and specialised uses such as the tiles on a space shuttle.
    3.1, 3.4
Modern ceramics containing Al_2O_3, SiC, Si_3N_4 have special uses.
    3.1.2, 3.4, 3.4.3, 3.4.9, 3.6.3

Recycling of glass   **3.12**

MATERIALS

2. CARBON-BASED POLYMERS

Structure
Carbon-based polymers have large molecules and high M_r. **4.2**
The structure is determined by covalent bonds and intermolecular forces.
Molecules may be linear or branched **4.3.2**
or cross-linked. **4.12**
See tacticity **4.3.2**

Polymerisation of a monomer or pair of monomers to a polymer can be by
- addition polymerisation, with a free radical mechanism, **4.6**
- condensation polymerisation. **4.12**

Manufacture
Addition polymerisation, **4.6, 4.6.1**
e.g. poly(phenylethene) **4.10, 4.13**
Condensation polymerisation **4.12, 4.13**
Fibres and fabrics **4.16.1, 4.16.2**
Regenerated fibres **4.18**

Uses
PVC **4.6**
poly(ethene) **4.8.2, 4.9.2**
isotactic poly(propene) **4.3.2**
fibres **4.16**
nylon **4.17.1, 4.22**
Terylene **4.17.3, 4.22**

Processing
Thermoplastics are shaped from molten plastic by injection moulding, extrusion, etc. **4.3.1, 4.15**
Thermosets are shaped during synthesis, and once set cannot be reshaped – except by machining. **4.3.1, 4.15**
Thermosets are formed *in situ* in **laminates**. **4.12.6, 4.15, 5.3.3**
Making a **fibre** **4.16.1**
Making a **fabric** **4.16.2**
Regenerated fibres, e.g. acetate, rayon. **4.18**
Rubbers are moulded. Vulcanisation increases the strength. **4.19, 4.19.2**
Modification The properties of polymers are improved by **copolymerisation** and by additives, e.g. stabilisers. **4.14, 4.19.3**

Deformation can be plastic, **1.4.1**
elastic, **4.19**
or viscoelastic **4.5.2**
and can arise from rotation of polymer chains. **4.3.2**

Properties
Most polymers are electrical and thermal insulators. **4.3**
In polymers **crystalline regions** show a melting temperature, while **amorphous regions** show a glass transition temperature. **4.3, 4.5**
The density depends on the degree of crystallinity. **4.9**
Cross-linked polymers are hard and brittle. **4.12.4–6**

Recycling of plastics **4.23**

Drawing plastics makes them stronger, e.g. the manufacture of fibres.
4.4, 4.5, 4.17, 4.17.1, 4.17.2, 4.17.4

3. METALS AND ALLOYS

Modification
The properties of metals are changed for specific uses by **alloying**. **2.4, 2.9.6, 2.13**
They may form **solid solutions** — interstitial or substitutional, **2.8** or **eutectics** **2.9, 2.9.6** for which **phase diagrams** can be drawn. **2.9.3**

Summaries
Mechanical properties of materials **6.2.1–4**
Structures of materials **6.2.7**
Deformation of materials **6.2.2**
Comparison of structural materials **6.4**
Stability of materials **6.6.1–3**
A comparison **6.6**
Recycling **2.15, 3.12, 4.23**

Processing
Some metals are shaped by **casting** molten metal. Nucleation plays a part in solidification. **2.11.1, 2.11.2**
Alloys have lower melting temperatures than pure metals. **2.9.6**
Metals can be shaped by
- **hot working,** **2.10.5, 2.11.4**
- **cold working.** **2.10.1, 2.11.3**

Annealing restores ductility in cold-worked metals. **2.7.3, 2.10, 2.10.4**

Metal surfaces need to be prepared for examination. **2.9.4, 2.9.5**

Corrosion
Metals tend to corrode, especially in the presence of electrolytes. There are methods of preventing corrosion.
2.12, 2.12.1, 2.12.2
A-level Chemistry 24.14.4, 24.14.5

Recycling of metals **2.15**

METALS

Structure
Metals owe their properties to the **metallic bond**. **2.2, 2.3, 2.3.3**
Metals are polycrystalline with a **grain structure**.
2.6, 2.6.1, 2.6.2, 2.6.7, 2.9.5, 6.2.7, 6.3.1
The **crystal structures**, bcc, fcc, hcp, have high coordination numbers. **2.5**
The **strength,** **1.2, 1.4–7** and **toughness,** **1.10, 1.11** of metals are determined by **dislocations** within grains. **2.7**

Properties
Metals have sharp melting temperatures, high densities, are electrical and thermal conductors, are lustrous, sonorous, ductile and malleable. **2.2**
They have high **tensile strength.** **1.4, 1.5**
Plasticity is due to slip and the movement of dislocations.
1.5, 1.6, 1.9, 2.7

MODULE 4823: MATERIALS

Note that the references in this module are to *Materials Science* by E. N. Ramsden (Stanley Thornes Publishers, 1995).

1. Ceramics

(a)(b) different types of ceramics:	
(i) simple crystal structures, complex silicate structures;	3.1.1, 3.5
(ii) amorphous ceramics;	3.1.1, 3.5.5
(iii) crystalline ceramics;	3.1.1
(iv) bonded ceramics;	3.1.1
(v) cements	3.1.1, 3.6.1, 3.6.2
(c)(d) bonding in ceramic materials	3.5
(d) the SiO_4^{2-} unit in silicates	3.5
(e)(f) the properties and uses of ceramics	3.1, 3.1.2, 3.4
(g) (i) the brittleness of ceramics and glasses caused by Griffiths cracks compared with the plasticity of metals; experiments	1.10, 3.1.2, 3.5.6
(h) industrial manufacture of cement	3.6
(i) experimental work	see lab notebook
(j)(k)(l) the formation of glasses, changes on cooling from the liquid state, glass transformation temperature	3.5.5, 4.5
(m) the addition of oxides to a glass	3.11
(n) heat treatment of glass	3.8
(o) devitrification	3.10
(p) glass ceramics	3.11.2
(q) ceramics containing aluminium oxide, silicon carbide, silicon nitride.	3.1.2, 3.4, 3.4.3, 3.4.9, 3.6.3

2. Carbon-based polymers

(a) (i) the terms 'monomer' and 'polymer'	4.2
(ii) addition and condensation polymerisation	4.6, 4.12
(b) chain length and molecular mass of polymers	4.2
(c) primary bonds (covalent) and secondary bonds (van der Waals' forces and hydrogen bonds) in polymer formation	4.2, 4.3.1, 4.3.2, 4.4, 4.17.1, 4.17.2
(d) the terms:	
linear, branched,	4.3.2
cross-linked,	4.12
co-polymer,	4.14
tacticity,	4.3.2
thermoplastic and thermoset	4.3.1
(e) deformation in terms of stress and rotation of chains,	4.3.2
plastic deformation,	1.4.1
elastic deformation,	4.19
viscoelastic deformation	4.5.2
(f) mechanism of free-radical addition polymerisation	4.6, 4.6.1
(g) the large-scale production of	
(i) poly(phenylethene)	4.6.1, 4.10, 4.13, 4.13.2
(ii) Terylene	4.12.1, 4.13
(iii) a nylon	4.12.3, 4.13
(iv) a phenol-methanal resin	4.12.4, 4.13
(h) predicting some physical properties of a polymer from its structure	Questions on Chapter 4
(i) the uses of polymers in relation to their structures	4.7, 4.8.2, 4.9.2, 4.10, 4.11, 4.12
(j)(k) cold drawing as applied to fibres and its effect on properties	4.17.1
(l) how physical properties are related to uses in nylon	4.17.1, 4.22
Terylene	4.17.3, 4.22
isotactic poly(propene)	4.3.2
(m) the manufacture of regenerated fibres	4.18
(n) the differences between natural and synthetic rubbers	4.19, 4.19.2, 4.19.3
(o) vulcanisation of rubber	4.19.2
(p) the use of additives	4.14
(q) uses of polymers related to properties and structures	4.1, 4.3, 4.4, 4.7, 4.8.2, 4.9.2, 4.11, 4.12, 4.14, 4.16–22

3. Metals and alloys

(a) strength, ductility and hardness	2.2, 2.3.3
(b) crystalline structure; grain structure	2.6.1, 2.6.2, 2.6.7, 2.7, 6.2.7
(c) description of as-cast structures of metals	2.9.4
(d) the role of nucleation in solidification	2.11.2
(e)–(h) X-ray diffraction studies of metals; crystal lattice; unit cell; hexagonal/face-centred close packing and body-centred cubic packing; coordination number and metallic radius related to type of structure	2.5

(i) interstitial solutions and substitutional solid solutions	**2.8**	(o) delocalisation of electrons in metals	**2.3** and *A-level Chemistry* **6.2.1**
(j) eutectics	**2.9, 2.9.6**		
(k) composition–temperature phase diagrams	**2.9.3**	(p)(q) plasticity of metals; slip; dislocations; dislocation tangling	**1.5, 1.6, 1.9, 2.7**
(l)(n) etching and examination of metal surfaces and appearance	**2.9.4, 2.9.5**	(r) strength of metals	**1.4, 1.5**
(m) polycrystalline nature of metals	**2.6, 2.6.1, 2.9.5, 6.3.1**	(s) strength of alloys	**2.4**

4824: ENVIRONMENTAL CHEMISTRY

Note that the references in this module are to *Chemistry of the Environment* by E. N. Ramsden (Stanley Thornes, 1996).

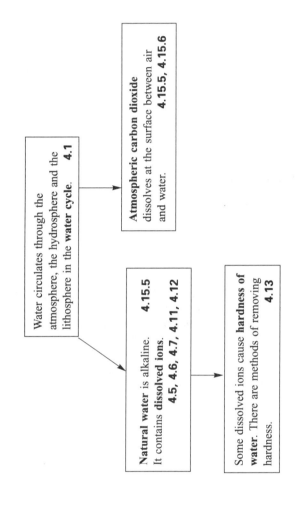

3. THE HYDROSPHERE

Water circulates through the atmosphere, the hydrosphere and the lithosphere in the **water cycle**. **4.1**

Atmospheric carbon dioxide dissolves at the surface between air and water. **4.15.5, 4.15.6**

Natural water is alkaline. **4.15.5** It contains **dissolved ions**. **4.5, 4.6, 4.7, 4.11, 4.12**

Some dissolved ions cause **hardness of water**. There are methods of removing hardness. **4.13**

1. THE ATMOSPHERE

The atmosphere consists of **troposphere, stratosphere, mesosphere** and the **mosphere**. **2.3**

In the stratosphere conditions favour free radical reactions. The concentration of ozone is maintained by a balance between such reactions. The **ozone layer** in the stratosphere reduces the intensity of UV radiation reaching Earth's surface. UV radiation is absorbed in the mesosphere and thermosphere. **2.4, 2.10, 3.11**

The **carbon cycle** balances photosynthesis, respiration and dissolution to maintain a constant level of carbon dioxide in the atmosphere. **2.8**

The **nitrogen cycle** balances processes which remove nitrogen from the atmosphere and processes which return it. **2.6**

The **residence time** of a substance in the atmosphere is governed by the relative rates of formation and removal. **2.4**

2. AIR POLLUTION

Photochemical reactions in the atmosphere involving NO and NO_2 lead to photochemical smog. **3.4, 3.8**

The level of **carbon dioxide** in the atmosphere is increasing. It contributes to **global warming**, together with other gases. **2.9, 3.11, 3.13**

Vehicle engines emit pollutants. Lean-burn engines and catalytic converters reduce the pollution. **3.9, 3.10**

Carbon monoxide is removed from the atmosphere by soil bacteria. **3.3**

Pollutant gases emitted into the atmosphere include CO_2 CO, SO_2, H_2S, NO, NO_2, NH_3, hydrocarbons and halogenoalkanes. **Particulates** include dust, smoke and lead compounds. **3.1–3.7**

Chlorofluorohydrocarbons, **CFCs** find many uses but they react with ozone and reduce the thickness of the ozone layer in the stratosphere. **3.11**

Oxides of sulphur are produced by the combustion of sulphur-containing fuels in power stations, factories and motor vehicles. They play a big part in the formation of **acid rain**. Methods of controlling SO_2 emission are in use. **3.4, 3.6**

In the lower atmosphere **ozone** is a pollutant. **3.12**

Radon, emitted from uranium-bearing rocks, is a radioactive hazard. **3.14**

4. WATER POLLUTION

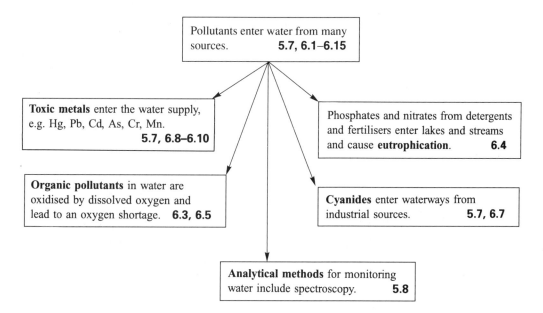

5. WATER SUPPLY AND SEWAGE TREATMENTS

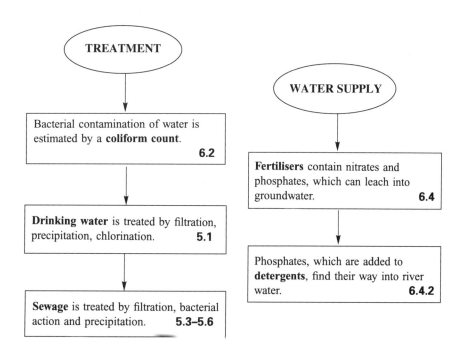

6. THE LITHOSPHERE

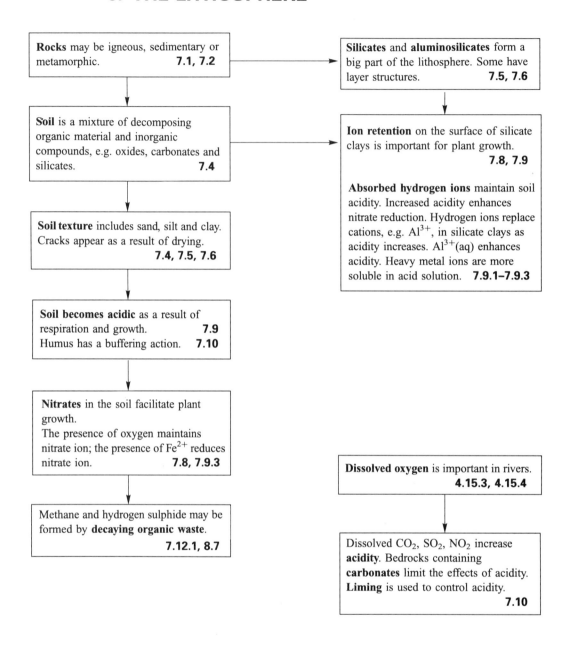

7. POLLUTION OF THE LITHOSPHERE

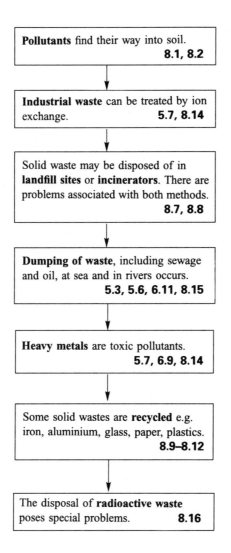

MODULE 4824: ENVIRONMENTAL CHEMISTRY

Note that the references in this module are to *Chemistry of the Environment* by E. N. Ramsden (Stanley Thornes, 1996).

1. The atmosphere

(a) the atmosphere: troposphere, stratosphere, mesosphere, thermosphere	2.3
(b) composition and temperature variation	2.3
(c) the carbon cycle: photosynthesis; respiration; carbon dioxide dissolved in surface water, calculation of $[CO_2(aq)]$ from data	2.8
(d) the nitrogen cycle: production of NO and NO_2 in lightning discharges; removal of NO_2 by solution in water	2.6

2. Air pollution

(a) emissions into the atmosphere	3.1–7
(b) the pollutants: carbon dioxide, carbon monoxide, hydrocarbons, halogenocompounds, sulphur dioxide, hydrogen sulphide, oxides of nitrogen, ammonia, and lead compounds; significance, sources, residence time, accumulation	3.1–7
(c) photochemical reactions	2.4, 3.8
(d)(e) atmospheric carbon dioxide concentration; seasonal variations; global warming	2.9, 3.11, 3.13
(f) IR absorption by dinitrogen oxide, methane, trichlorofluoromethane and global temperature	2.4, 2.10, 3.11
(g)(h) removal of carbon monoxide from the atmosphere	3.3

3. The hydrosphere

(a)(b) describe and explain the water cycle	4.1
(c) the causes of hardness of water	4.13
(d) methods for removing the hardness of water	4.13
(e) equilibria which make natural water alkaline	4.15.5

4. Water pollution

(a) types of water pollutants	6.1–6.15
(b) oxygen demand	4.15.3–4, 6.3
(c) movement of pollutants in the hydrosphere	6.15
(d) the major trace metals in water (Hg, Pb, Cd, As, Cr, Mn)	6.8, 6.9, 6.10

5. Water supply and sewage treatment

(a) bacterial contamination: the coliform count	6.2
(b) the preparation of potable water	5.1
(c) the treatment of sewage	5.3
(d) chlorinated organic matter in domestic water	5.1, 7.7

(e)(f) the stratospheric concentration of ozone as maintained by:	
(i) the photodissociation of NO_2, O_2 and O_3 to give reactive oxygen atoms;	2.10
(ii) the formation of O_3 and OH by reaction of oxygen atoms with O_2 and H_2O;	2.10
(iii) the reaction between O_3 and NO	2.10
(g) the effect of ozone on the intensity of UV radiation at the Earth's surface	2.7, 2.10
(h) 'residence time'	2.4
(i) absorption of far ultraviolet in the mesosphere and thermosphere	2.3, 2.4

(i)(j) sources of sulphur dioxide in the atmosphere; acid rain	3.4
(k) effects of sulphur dioxide on people, plants, materials	3.4
(l) industrial methods of controlling sulphur dioxide emission	3.6
(m) sources of nitrogen oxides in the atmosphere, photochemical smog and its effects	3.5, 3.8
(n) the pollution effects of the internal combustion engine, lean-burn engines and catalytic converters	3.9
(o) effects of ozone in the lower atmosphere	3.12
(p)(q) the effect of CFCs on the ozone layer; possible alternatives	3.7.3
(r) radon emission	3.14

(f) the sources of dissolved ions in natural water	4.5–7, 4.11, 4.12
(g) the air/water interface: temperature and partial pressure and their effects on the solubility of carbon dioxide	4.15.5–6

(e) analysis of water (e.g. AAS)	5.8
(f)(g)(i) phosphates, nitrates, cyanides in waterways	6.4, 6.7
(h) eutrophication	6.4

(e) nitrogen- and phosphate-containing fertilisers	6.4
(f) the leaching of nitrate from fertilised land	6.4, 6.4.1
(g) use of phosphates in detergents; disadvantages	6.4.2

ENVIRONMENTAL CHEMISTRY

6. The lithosphere

(a) igneous, sedimentary and metamorphic rocks	7.1, 7.2
(b) silicates and aluminosilicates	7.6, 7.8
(c)(d) soil as a mixture of organic and inorganic compounds	7.4, 7.7
(e) containing oxides, carbonates and silicates	7.5, 7.6, 7.10
(f)(g)(h) the structure of layer silicates, 1 : 1 and 2 : 1 layer silicates, ion substitution within layer silicates	7.5, 7.6, 7.8
(i) the formation of cracks in soils as a result of drying	7.6
(j) ion retention on the surface of silicate clays	7.5, 7.6, 7.8
(k)(l)(m) the role of adsorbed hydrogen ions in maintaining soil acidity and enhancing nitrate reduction	7.9
(n) release of aluminium ions by silicate clays in acid solution	7.9.2
(o) the role of oxygen in maintaining nitrogen in compounds which facilitate plant growth	7.8, 7.9.3
(p) the effects of reducing conditions on the reduction of nitrate ions	7.8, 7.9.3
(q) and on the formation of methane and hydrogen sulphide	7.12.1, 8.7
(r) dissolved oxygen in rivers	4.15.3, 4.15.4
(s) the effect of oxides of carbon, sulphur and nitrogen on the pH of water	3.6, 4.1
(t) soil acidification due to respiration and growth	7.9
(u) how the hydrated aluminium ion enhances acidity	7.9.2
(v) the increased solubility of heavy metal ions in acid solution	7.9.3
(w) how bedrocks containing carbonates counteract acidity	7.10
(x) the buffering actions of the HCO_3^- ion and humus	7.10
(y) the role and limitations of liming in acidity control	7.10

7. Pollution of the lithosphere

(a) soil pollution	8.2
(b) ion exchange in the treatment of industrial waste	5.7, 8.14
(c) land-filling and incineration for the disposal of solid waste	8.7, 8.8
(d) dumping of waste at sea and in rivers	5.3, 5.6, 6.11, 8.15
(e) heavy metals in the environment	3.2, 6.9, 4.7, 5.3, 5.6, 8.7, 8.14
(f) the advantages and difficulties of recycling waste materials	8.9–8.12
(g) disposal of radioactive waste	8.16

4825: METHODS OF ANALYSIS AND DETECTION

Note that the references in this module are to *Detection and Analysis* by E. N. Ramsden (Stanley Thornes, 1996).

1. SEPARATION TECHNIQUES FOR ANALYSIS: CHROMATOGRAPHY AND ELECTROPHORESIS

Chromatography: The components of a mixture can be separated by:
1. methods which depend on **adsorption**
   - column chromatography **1.2**
   - thin layer chromatography **1.4**
2. methods which depend on **partition**
   - paper chromatography **1.3**
   - gas chromatography **1.5**
   - liquid chromatography **1.6**
3. methods which depend on **size**
   - gel permeation chromatography **1.8**
4. ion exchange **1.7**

Electrophoresis is used to separate the components of a mixture on the basis of either the speed at which they move in an electric field **2.1**
or particle size **2.2**
Uses include the separation of proteins and genetic fingerprinting. **2.3**

2. MASS SPECTROMETRY

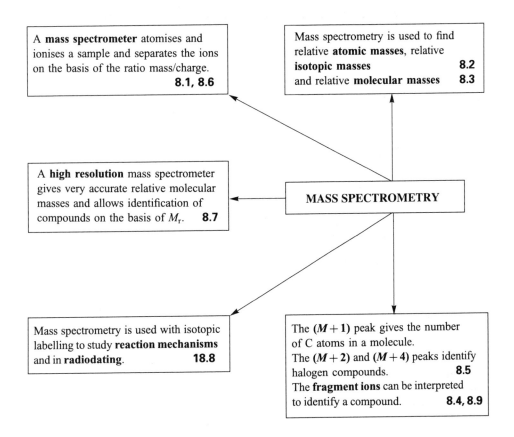

3. SPECTROSCOPIC METHODS OF ANALYSIS

3.1 BASIC PRINCIPLES

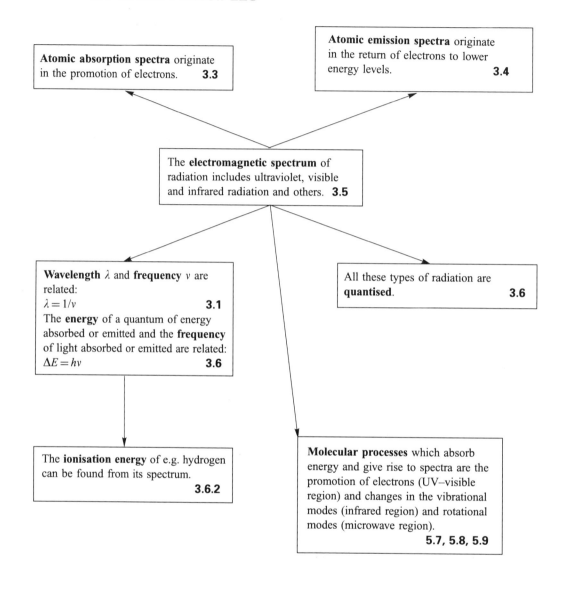

3.2 ATOMIC SPECTROSCOPY

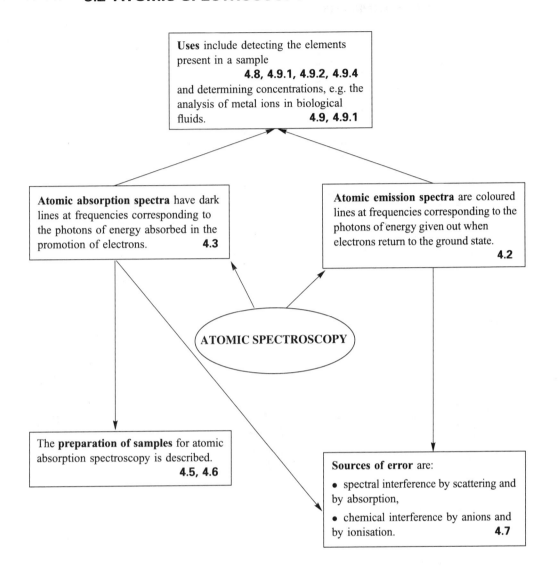

3.3 ULTRAVIOLET AND VISIBLE SPECTROSCOPY INCLUDING COLORIMETRY

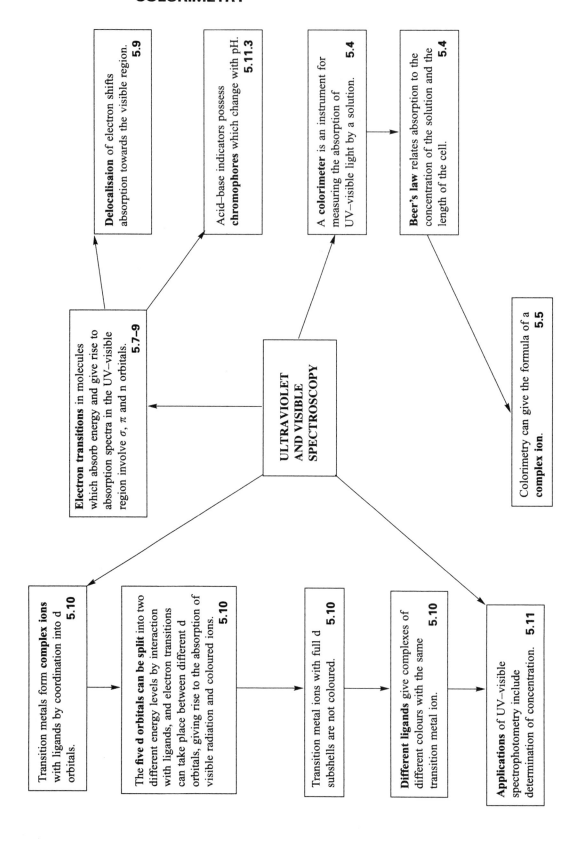

3.4 INFRARED SPECTROSCOPY

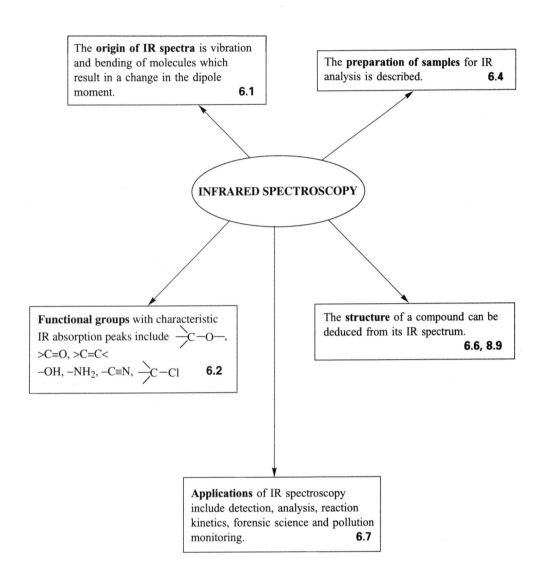

3.5 NUCLEAR MAGNETIC RESONANCE SPECTROSCOPY

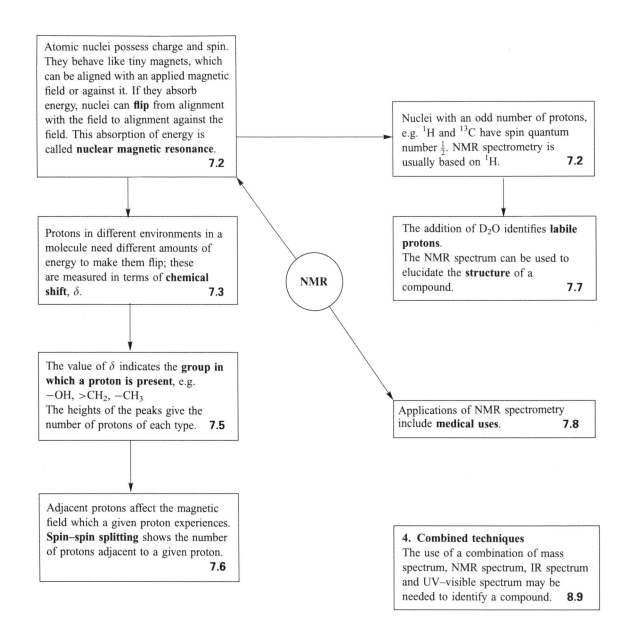

MODULE 4825: METHODS OF ANALYSIS AND DETECTION

Note that the references in this module are to *Detection and Analysis* by E. N. Ramsden (Stanley Thornes, 1996).

1. Separation techniques for analysis: chromatography and electrophoresis

(a) chromatography: paper, high performance liquid, thin layer and gas/liquid	1.1–5
(b)(c) interpretation of chromatograms: paper, TLC and gas/liquid	1.3–6
(d) electrophoresis	2.1, 2.2
(e) hydrolysis of proteins and separation of the products by electrophoresis	2.3
(f) applications in industry and medicine	1.5, 2.3.1–3
(g) analysis of genes; genetic fingerprinting	2.3.3

2. Mass spectrometry

(a) basic principles	8.1
(b) high resolution mass spectrometry	8.7
(c) the use of the $(M+1)$ peak to give the number of carbon atoms per molecule	8.5
(d) the use of the $(M+2)$ and $(M+4)$ peaks in the identification of halogen compounds	8.5
(e)(f) identifying the major fragment ions in a mass spectrum	8.4, 8.5
(g) the use of isotopic labelling to determine the position of reaction in a molecule and of radiodating to determine age	8.8.7, 8.8.8
(h) use of peaks to elucidate structure	8.8.2

3. Spectroscopic methods of analysis

3.1 Basic principles

(a) the electromagnetic spectrum	3.5
(b) absorption and emission atomic spectra	3.3, 3.4
(c) quantisation of electromagnetic radiation	3.6
(d) relationship of wavelength to frequency and energy change to frequency	3.1, 3.2
(e) ionisation energy predicted from convergence of atomic line spectra	3.6.2
(f) the molecular processes which bring about absorption of energy	5.1, 5.2, 5.7, 5.8, 5.9

3.2 Atomic spectroscopy

(a) principles of atomic absorption and emission spectra	4.1–4
(b) preparation of samples for atomic absorption spectroscopy	4.5, 4.6
(c) sources of interference in atomic spectra	4.7, 4.8
(d) ions present in a sample deduced from its spectrum	4.8, 4.9
(e) concentration of ions in a sample found from its spectrum	4.8, 4.9.1
(f) uses of emission spectra in analysis	4.9

3.3 Ultraviolet and visible spectroscopy

(a) action of e.g. NH_3, H_2O, Cl^- as ligands	5.10
(b) the splitting of degenerate d orbitals in octahedral complexes	5.10
(c) the effects of different ligands on the colour of a given transition metal complex	5.10
(d) the colour of a complex predicted from its visible spectrum	2.2, 5.3, 5.7, 5.9
(e) the lack of colour in complexes of Zn^{2+}, Pb^{2+}, Cu^{2+} and in TiO_2	5.10
(f) the electronic transitions responsible for absorption in organic molecules	5.1, 5.2, 5.7–9
(g) the effects of delocalisation on absorption in the visible region	5.9
(h) the colour changes in acid-base indicators in terms of a change in the chromophores	5.11.3
(i)(j) colorimetry; use of Beer's Law, $\lg(I_0/I) = Ecl$ to find concentration	5.4
(k) absorbance experiments to determine concentrations	5.4, 5.11
(l)(m) determination of stoichiometry of complex ion and calculations	5.5

3.4 Infrared spectroscopy

(a) IR absorptions for simple molecules and the molecular vibrations which give rise to them	6.1
(b) absorptions due to functional groups	6.2
(c) preparation of samples of liquids and solids for IR analysis	6.4
(d) relating the IR spectrum to the structure of a compound containing up to three functional groups	6.6, 8.9
(e) IR spectroscopy in analysis, e.g. forensic science, air pollution	6.7

3.5 Nuclear magnetic resonance spectroscopy

(a) the principles of nuclear magnetic resonance in ^1H and ^{13}C	7.2
(b) how the chemical environment of a proton affects the absorption of energy at resonance; use of the δ scale, and TMS as a standard	7.3
(c) the effects of adjacent protons on the magnetic field experienced by a given proton	7.6
(d) integration of peak area to give the number of protons in each group	7.5
(e) use of spin–spin splitting to give the number of protons adjacent to a given proton	7.6

(f) structure of a compound containing up to three functional groups	7.7	(h) medical uses of NMR spectroscopy	7.8
(g) the addition of D_2O to identify labile protons	7.7		

4. Combined techniques

(a)(b)(c) the use of several spectra to identify an unknown compound	**8.9** and **Questions on Combined Techniques**

4826: HOW FAR, HOW FAST?

1. CHEMICAL ENERGETICS

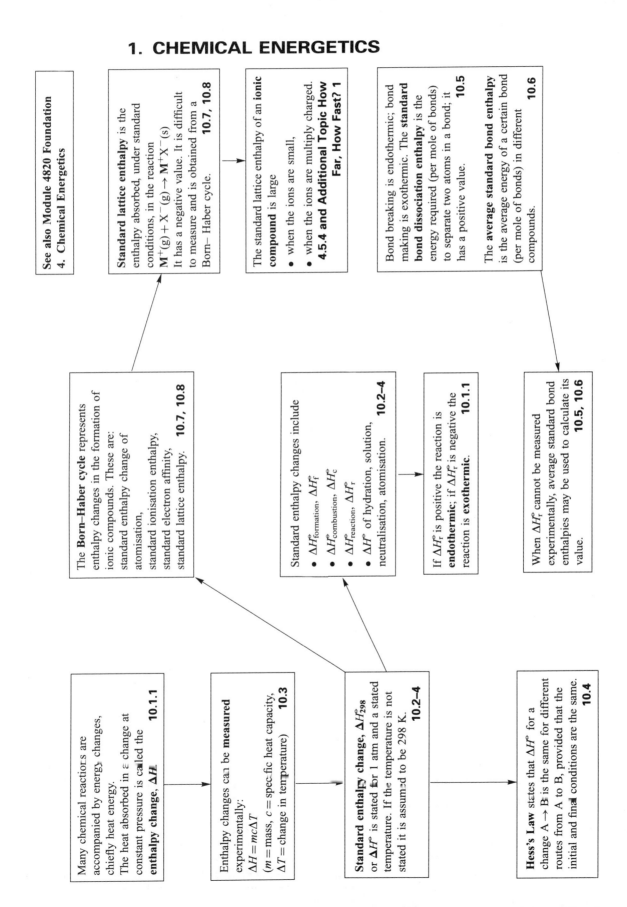

See also Module 4820 Foundation 4. Chemical Energetics

Standard lattice enthalpy is the enthalpy absorbed, under standard conditions, in the reaction
$M^+(g) + X^-(g) \rightarrow M^+X^-(s)$
It has a negative value. It is difficult to measure and is obtained from a Born–Haber cycle. **10.7, 10.8**

The standard lattice enthalpy of an **ionic compound** is large
- when the ions are small,
- when the ions are multiply charged.
4.5.4 and Additional Topic How Far, How Fast? 1

Bond breaking is endothermic; bond making is exothermic. The **standard bond dissociation enthalpy** is the energy required (per mole of bonds) to separate two atoms in a bond; it has a positive value. **10.5**

The **average standard bond enthalpy** is the average energy of a certain bond (per mole of bonds) in different compounds. **10.6**

The **Born–Haber cycle** represents enthalpy changes in the formation of ionic compounds. These are: standard enthalpy change of atomisation, standard ionisation enthalpy, standard electron affinity, standard lattice enthalpy. **10.7, 10.8**

Standard enthalpy changes include
- $\Delta H^\ominus_{formation}$: $\Delta H^\ominus_f$
- $\Delta H^\ominus_{combustion}$: $\Delta H^\ominus_c$
- $\Delta H^\ominus_{reaction}$: $\Delta H^\ominus_r$
- $\Delta H^\ominus$ of hydration, solution, neutralisation, atomisation. **10.2–4**

If $\Delta H^\ominus_r$ is positive the reaction is **endothermic**; if $\Delta H^\ominus_r$ is negative the reaction is **exothermic**. **10.1.1**

When $\Delta H^\ominus_f$ cannot be measured experimentally, average standard bond enthalpies may be used to calculate its value. **10.5, 10.6**

Many chemical reactions are accompanied by energy changes, chiefly heat energy. The heat absorbed in a change at constant pressure is called the **enthalpy change, ΔH**. **10.1.1**

Enthalpy changes can be **measured** experimentally:
$\Delta H = mc\Delta T$
(m = mass, c = specific heat capacity, ΔT = change in temperature) **10.3**

Standard enthalpy change, $\Delta H^\ominus_{298}$ or $\Delta H^\ominus$ is stated for 1 atm and a stated temperature. If the temperature is not stated it is assumed to be 298 K. **10.2–4**

Hess's Law states that $\Delta H^\ominus$ for a change A → B is the same for different routes from A to B, provided that the initial and final conditions are the same. **10.4**

2. ELECTROCHEMISTRY

Cells
Two electrodes of different $E^\ominus$ values combine to form a cell. The **standard cell potential** is given by
$E^\ominus_{cell} = E^\ominus_{RHS\ electrode} - E^\ominus_{LHS\ electrode}$
A positive value of $E^\ominus_{cell}$ indicates that the cell reaction will happen spontaneously and that electrons will flow from left to right through the external circuit. **13.1**

The **equation for a redox reaction** is obtained by adding **half-equations** for the changes in the oxidising agent and the reducing agent. **3.15**

In a redox reaction, sum of increases in oxidation number of oxidised species = sum of decreases in oxidation number of reduced species, e.g. $2Fe^{3+} + 2I^- \rightarrow 2Fe^{3+} + I_2$ **3.16.2, 3.16.3**

Listing oxidants and reductants in order of $E^\ominus$ values gives a **redox series**. **12.2**

The $E^\ominus$ values of two substances predict the **direction of any redox reaction** between them. **13.1.2**

Oxidation involves an increase in **oxidation number**; reduction involves a decrease in oxidation number. **3.15–17**

The **standard electrode potential** of ions of the same electrode in different oxidation states can be found. **Additional Topic How Far, How Fast? 2**

For a **redox system**,
Oxidant + $ne^- \rightleftharpoons$ Reductant
An electrode immersed in the system acquires a potential that depends on the position of the equilibrium between the oxidant and reductant. **3.15–17**

The **standard electrode potential** $E^\ominus$ is the potential of an electrode in a 1 mol dm^{-3} solution of ions at 298 K. **13.1, 13.1.1**

An electrode potential is measured by finding the emf of a cell composed of the electrode and a standard electrode, e.g. the **standard hydrogen electrode**, which has $E^\ominus = 0$ by definition or the calomel electrode. **13.1.2**

The value of electrode potential depends on $E^\ominus$ and also on the **concentration of ions** in solution.
Additional Topic How Far, How Fast? 3

HOW FAR, HOW FAST?

3. REACTION KINETICS

Catalysts provide an alternative reaction route with a lower activation energy and a higher rate constant. **14.9, 14.10, 14.15**
Enzymes are biological catalysts.
Additional Topic How Far, How Fast? 5
Buffering is important for enzyme catalysis.
Additional Topic How Far, How Fast? 6

A **homogeneous catalyst** is in the same phase as the reactants, e.g.
- transition metal compounds,
- enzymes. **14.10.1, 25.9.2, 32.11**

A **heterogeneous catalyst** is in a different phase from the reactants, e.g. the transition metals, iron in the Haber Process and platinum in catalytic converters in vehicles **14.10.2**

The **Maxwell–Boltzmann distribution** is a graph of the fraction of a set of molecules that has a certain energy against that energy. Graphs at different temperatures show that, as well the average energy increasing, the fraction of molecules with high energy increases with temperature. **7.3.4, 14.9.1, 14.10**

The rate constant increases with **temperature**. From the relationship
$$k = Ae^{-E/RT}$$
the quantity E, the **activation energy** of the reaction, can be found. Colliding particles must have this amount of energy in order to react. **14.9.1, 14.9.2**

For a reaction, $A \rightarrow X$
Initial rate of reaction $= k[A]_0^m$
where $[A]_0 =$ initial concentration of A and $m =$ order with respect to A. Comparing initial rates at different $[A]_0$ values gives the order, m. **14.5.1**

Experimental techniques for following the course of a reaction include
- titrimetric analysis,
- spectrophotometric analysis,
- measurement of the pressure of a gaseous reactant,
- measurement of electric potential or conductance. **14.1.1–5, 14.3**

The **rate equation** for a chemical reaction is derived from measurements of the concentrations of reactants or products after measured intervals of time. **14.2**
Rate $= k[A]^m[B]^n$
where [A] and [B] are concentrations of A and B, $m =$ **order of reaction** with respect to A, $n =$ order with respect to B, $k =$ **rate constant**. **14.4.2**

Graphs of [reactant] or [product] against time elapsed since the start of the reaction have different shapes for zero-, first- and second-order reactions. **14.5.2, 14.5.3, 14.6**
First order reactions have a **half-life** which is independent of [reactant]. **14.5.3**

The **rate of a chemical reaction** (change in concentration/time) is influenced by:
- the concentrations of reactants in solution, **14.1.2, 14.4**
- the pressures of gaseous reactants, **14.1.2**
- the surface areas of solid reactants, **14.1.1, 14.4**
- temperature, **14.1.3, 14.8**
- light, **14.1.4, 14.7**
- catalysts. **14.1.5, 14.1.6, 14.10**

Many reactions take place in a number of steps. In many cases only one step is the **rate-determining step**. **14.1**
Many reactions proceed via a **transition state**. **14.9.2**
The results of kinetic experiments give evidence for the mechanism of a chemical reaction.
14.11, 14.12 and Additional Topic How Far, How Fast? 4

4 EQUILIBRIA

Many chemical reactions are **reversible**:
$A \rightleftharpoons B$

A reversible reaction may reach a state of equilibrium. In a **dynamic equilibrium** the forward and reverse reactions are still happening. The extent to which the position of equilibrium favours the products over the reactants is measured by the **equilibrium constant**.   3.19, 3.20, 11.1, 11.2

Le Chatelier's Principle predicts the direction in which the equilibrium will shift as a result of changes in **concentration** and **pressure**.   11.4

Applications of Le Chatelier's Principle include

- the manufacture of **ammonia** in the Haber Process,   11.5.3, 22.4.1
- the oxidation of ammonia to nitric acid,   22.7.3
(see the importance of ammonia and nitrogen compounds)
  22.2.1, 22.4.1, 22.7.3
- the manufacture of **sulphuric acid** in the Contact Process   21.11

(see the importance of sulphuric acid).
  21.11

For the equilibrium
$A(g) + B(g) \rightleftharpoons C(g) + D(g)$
the equilibrium constant K_p is:
$K_p = \dfrac{p_C \times p_D}{p_A \times p_B}$
p_C = partial pressure of C = mole fraction of C × total pressure
  3.21, 11.1–5

For the equilibrium
$E(aq) + F(aq) \rightleftharpoons G(aq) + H(aq)$
the equilibrium constant K_c is:
$K_c = [G][H]/[E][F]$
where [G] = molar concentration of G
  11.1–3, 11.5

Changes in **temperature** affect the equilibrium constant.   11.4

The presence of a **catalyst** does not affect the position of equilibrium but only the time taken to reach equilibrium.   11.4

HOW FAR, HOW FAST?

Water is slightly dissociated:
$H_2O(l) \rightleftharpoons H^+(aq) + OH^-(aq)$
The **ionic product** for water,
$K_w = [H^+(aq)][OH^-(aq)] = 1.0 \times 10^{-14} \text{ mol}^2 \text{ dm}^{-6}$ at 298 K
12.7.2

Hydrogen ion concentration is expressed as **pH**:
$pH = -\log[H^+(aq)/\text{mol dm}^{-3}]$
12.7.3

For a **weak acid**,
$HA + H_2O \rightleftharpoons H_3O^+ + A^-$
$K_a = [H_3O^+][A^-]/[HA]$
K_a is the **dissociation constant** of the weak acid.
12.7.4

Buffer solutions
A solution of a weak acid and its salt and a solution of a weak base and its salt act as buffers. They absorb small amounts of hydrogen ion and hydroxide ion with only a small change in pH. **Amino acids** have a buffering action. The **pH of a buffer** consisting of a weak acid of dissociation constant K_a and its salt is given by
$pH = pK_a + \log([\text{Salt}]/[\text{Acid}])$
12.7.12

For a salt **AB** of very low solubility, the product of the concentrations of its ions is called the **solubility product**, K_{sp}.
$K_{sp} = [A^{2+}][B^{2-}]$
12.7.14

The solubility of **AB** is reduced by adding A^{2+}(aq) or B^{2-}(aq). This effect is called the **common ion effect**.
12.7.15

According to the Brønsted-Lowry theory, **acids** are proton-donors, **bases** are proton-acceptors. Acid-base conjugate pairs are: $AH + B \rightleftharpoons A + BH$
acid + base $\rightleftharpoons$ conjugate + conjugate base of **A** acid of **B**
12.7.1

Strong acids and strong bases are completely dissociated.
Weak acids and weak bases are only partially dissociated.
12.7.4

When an aqueous acid is titrated against an aqueous base, a plot of pH against volume of titrant can be plotted – a **titration curve**. The pH changes rapidly at the end point.
An **indicator** which changes colour at the end point should be chosen.
12.7.8, 12.7.9

MODULE 4826: HOW FAR, HOW FAST?

1. Chemical energetics		(c) Hess' Law used in calculations on	
(a) (i) enthalpy change of reaction and standard conditions	10.2, 10.3, 10.4	(i) the formation of a simple ionic solid and of its aqueous solution	10.7, 10.8
(ii) lattice energy	10.7, 10.8	(ii) average bond energies	10.5, 10.6
(b) the effects of ionic charge and ionic radius on the magnitude of lattice energy	4.5.4 and see **Additional Topic How Far, How Fast? 1**	(d) ΔH as a guide to feasibility of reaction but not rate of reaction	10.10

2. Electrochemistry		(e) standard cell potential used to explain/predict the direction of electron flow in a simple cell and to predict the feasibility of a reaction	13.1.2
(a) standard electrode potential and standard cell potential	13.1, 13.1.1		
(b) the standard hydrogen electrode	13.1.2		
(c) measurement of standard electrode potentials of metals or non-metals in contact with their ions in aqueous solution and of ions of the same element in different oxidation states	13.1.2	(f) kinetic factors that may prevent a feasible reaction from occurring	10.10
		(g) redox equations	3.15–17
		(h) variation of electrode potential with concentration	**Additional Topic How Far, How Fast? 2**
(d) calculation of standard cell potential	13.1.2		

3. Reaction kinetics		(d) calculation of rate constant from initial rate	**Check-points 14B, 14C, 14D**
(a) rate of reaction,	14.1, 14.1.6, 14.2		
rate equation,	14.2	(e) the effect of concentration changes on the rate of a reaction	14.4.2, 14.9.1
order of reaction,	14.4.2, 14.5		
rate constant,	14.4.2	(f) Boltzmann distribution	7.3.4
half-life of a reaction,	14.5.3	(g) activation energy, collision frequency, the effect of temperature change on a rate constant	14.9.1
rate determining step,	14.1		
activation energy,	14.9.2		
catalysis	14.1.5, 14.10	(h) the effect of a catalyst on rate constant, interpreted in terms of the Boltzmann distribution	14.9, 14.10
(b) rate equations of the form rate $= k[A]^m[B]^n$ (m and $n = 0$, 1 or 2)		(i) homogeneous and heterogeneous catalysis	14.10.1, 14.10.2
(i) order of reaction by the initial rates method	14.5.1 and **Checkpoint 14B**	(j) enzymes: biological catalysts with specific activity	14.10.1, 14.10.2 and **Additional Topic How Far, How Fast? 4**
(ii) order of reaction from graphs of concentration–time and rate–concentration	14.5.2, 14.5.3 and **Checkpoint 14C**		
(iii)(iv)(v) checking or proposing a reaction mechanism that is consistent with the kinetics and predicting order of reaction from mechanism	14.11, 14.12 and see **Additional Topic How Far, How Fast? 3**	(k) catalysts in the Haber process,	22.4.1
		petroleum processing,	26.3.2–26.3.4
		the use of immobilised enzymes in industry	See **Additional Topic How Far, How Fast? 4**
(vi) calculating an initial rate from concentration data	14.2, 14.5.1	(l) the control of emission from cars using catalytic converters	26.4
(c) half-life of a first-order reaction; calculations	14.5.3, **Checkpoint 14C**		

HOW FAR, HOW FAST?

4. Equilibria

(a) reversible reaction and dynamic equilibrium	3.19, 3.20, 11.1, 11.2
(b)(c) state and apply Le Chatelier's Principle	11.4
(d) equilibrium constants K_c and K_p	3.21, 11.1, 11.2, 11.3
(e)(f) calculation of equilibrium constants; calculation of equilibrium quantities	11.5
(g) the Haber process	11.5.3, 22.4.1
(h) the Contact process	21.11
(i) importance of ammonia and compounds made from it	2.4.1, Figure 22.1
(j) the importance of sulphuric acid	Figure 21.17
(k) the Brönsted–Lowry theory of acids and bases	12.7.1
(l) strong and weak acids and bases	12.7.4
(m)(n) the terms of pH, K_a, pK_a and K_w; calculations	12.7.2–12.7.4, 12.7.6
(Note that UCLES does not require the calculation of the pH of a solution of a weak base.)	
(o) suitable indicators for acid–base titrations	12.7.8
(p) change in pH during acid–base titration	12.7.9
(q)(r) buffer solutions: mode of action; calculation of pH	12.7.12 and see Additional Topic How Far, How Fast? 5
(s)(t) solubility product, K_{sp}; calculations	12.7.14
(u) the common ion effect	12.7.15

ADDITIONAL TOPIC HOW FAR, HOW FAST? 1: THE EFFECT OF IONIC CHARGE AND IONIC RADIUS ON LATTICE ENERGY

(Read after *A-level Chemistry* **4.5.3**)

The lattice energy of an ionic compound can be found by experiment ... ... and calculated from a theoretical model of the ionic bond.

The value of the **lattice energy** of an ionic compound is a measure of the stability of ionic crystals. An experimental value can be found by means of a Born–Haber cycle (**4.2.9**). You will find a fuller treatment of lattice energy under its more precise name of lattice enthalpy in **10.7**.

A simple model of an ionic bond is two point charges separated by a distance equal to the sum of the radii of the cation and anion. Energy is required to move the two charges further apart against the attraction between them. The model can be extended to a three-dimensional lattice of point charges, e.g. a crystal. The energy required to separate the charges, the lattice energy, can be calculated from the theoretical model. The resulting equation has the form:

$$\text{Lattice energy} \propto \frac{(\text{Number of charges on cation} \times \text{Number of charges on anion})}{(\text{Radius of cation} + \text{Radius of anion})}$$

Calculated values of lattice energies agree well with experimental values from the Born–Haber cycle for many compounds, e.g. the alkali metal halides. The agreement supports the model of an ionic compound and the assumption that the lattice energy depends on the radii and charges of the ions present.

SIZE OF IONS

Theory predicts that lattice energy $\propto$ (charge on cation $\times$ charge on anion) ... ... and $\propto 1/$(radius of cation + radius of anion).

The lattice energy is proportional to the reciprocal of the distance between ions, $1/(r_{\text{cation}} + r_{\text{anion}})$. For any given cation, the lattice energy of its compounds decreases as the size of the anion increases. Note the trend in lithium halides from F through Cl and Br to I in Table 1. For a given anion, the lattice energy of its compounds decreases as the size of the cation increases. Note the trend in iodides from Li through Na and K to Rb. For large cations, the lattice enthalpy of its compounds is determined mainly by the size of the cation and varies less with changes in the anion. Note the trend in rubidium halides compared with that for lithium halides.

TABLE 1 Lattice Energies/kJ mol^{-1} (experimental values)

Ion	Li$^+$	Na$^+$	K$^+$	Rb$^+$	Mg^{2+}
F$^-$	1031	918	817	783	2957
Cl$^-$	848	780	711	685	2526
Br$^-$	803	742	679	656	2440
I$^-$	759	705	651	628	2327
O^{2-}	2814	2478	2232	2161	3791

The table gives experimental values.

The lattice energy is higher for compounds with small ions ...
... because they can approach closely ...
... and for compounds with ions of multiple charge ...
... because electrostatic attraction between them is great.

CHARGES ON IONS

The lattice energy is proportional to the product of the charges on the cation and the anion. The ions Na$^+$ and Mg^{2+} are similar in radius. Compare the values of lattice energy for NaF and MgF$_2$. The ions F$^-$ and O^{2-} are similar in radius. Compare the values of lattice energy for Na$_2$O and MgO. Although both pairs of ions, Na$^+$ and Mg^{2+} and also F$^-$ and O^{2-} are similar in size, the lattice energy of MgO (3791 kJ mol^{-1}) is much greater than that of NaF (918 kJ mol^{-1}).

A compound of elements which do not differ greatly in electronegativity has some covalent character ...
... and the lattice energy is high.

ELECTRONEGATIVITY

For some compounds, e.g. silver halides and zinc sulphide, experimental values of lattice energies are greater than the theoretical values. The explanation put forward is that the bonding is not purely ionic. There is an incomplete transfer of valence electrons from cation to anion. As a result, there is some electron density in between the cation and the anion; that is, the bond has partial covalent character. This happens when the difference in electronegativity between the bonded atoms is not very great. The discussion of bond type continues in **4.5.4**.

CHECKPOINT ON ADDITIONAL TOPIC HOW FAR, HOW FAST? 1

1. Comment on the difference in lattice energy between (*a*) LiF and KF (*b*) NaF and NaI (*c*) NaCl and Na$_2$O.

2. Which compound in the table has the highest lattice energy? Suggest why.

ADDITIONAL TOPIC HOW FAR, HOW FAST? 2: METHOD OF MEASURING THE STANDARD ELECTRODE POTENTIAL OF IONS OF THE SAME ELEMENT IN DIFFERENT OXIDATION STATES

The diagram shows an apparatus which can be used to measure the standard electrode potential of an element in different oxidation states, e.g. Fe^{3+}/Fe^{2+}. The Fe^{3+}/Fe^{2+} system is combined with a saturated calomel electrode. A saturated calomel electrode has $E = 0.245$ V. In this cell,

$$E_{cell} = E_{Fe^{3+}/Fe^{2+}} - E_{calomel(sat)}$$
$$0.526 \text{ V} = E_{Fe^{3+}/Fe^{2+}} - 0.245 \text{ V}$$
$$E_{Fe^{3+}/Fe^{2+}} = 0.771 \text{ V}$$

FIGURE 1 Apparatus for Measuring the Standard Electrode Potential for an Element in Different Oxidation States

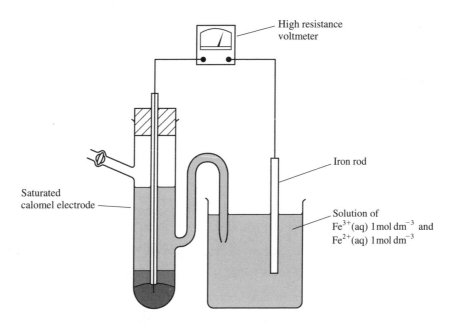

ADDITIONAL TOPIC HOW FAR, HOW FAST? 3: THE EFFECT OF CONCENTRATION ON ELECTRODE POTENTIAL

(Read after *A-level Chemistry* **13.1.3**)

The value of the electrode potential E of a metal depends ...

The value of the electrode potential of a metal depends on the standard electrode potential of the metal (see Table 13.1), on the temperature and also on the concentration of the metal ions in solution. It is standard practice to quote values at 298 K (25 °C). Let us look at the effect of concentration on electrode potential.

... on the standard electrode potential $E^{\ominus}$ ...
... on the temperature ...

In the system which consists of a silver electrode in equilibrium with a solution of silver ions,

$$Ag^+(aq) + e^- \rightleftharpoons Ag(s)$$

Le Chatelier's Principle can predict what will happen to the electrode potential if the concentration of silver ions is changed. If the solution is made more dilute, more silver atoms will ionise in an attempt to restore the equilibrium concentration of silver ions, more electrons will accumulate on the metal, and the electrode potential will become more negative (will have a larger negative value or a smaller positive value). On the other hand, if the concentration of metal ions is increased, some ions will take electrons from the electrode to form silver atoms, and the electrode potential will become less negative (have a smaller negative value or a larger positive value).

... and on the concentration of metal ions in solution.

The prediction can be tested by measuring the emf of a silver-copper cell,

$$Cu(s) \mid Cu^{2+}(aq) \mid Ag^+(aq) \mid Ag(s)$$

The concentration of copper ions is kept constant at 1.00 mol dm^{-3} and the concentration of silver ions is varied between 1.00 mol dm^{-3} and 1.00×10^{-4} mol dm^{-3}. The emf of the cell is given by

$$\text{Emf (cell)} = E(\text{Ag electrode}) - E(\text{Cu electrode})$$

The electrode potential increases linearly with the concentration of metal ions.

Therefore by subtracting E(Cu electrode) = 0.34 V from the emf of the cell, values of E(Ag electrode) can be found. A plot of E(Ag electrode) against lg[Ag$^+$(aq)], is shown in Figure 2.

The intercept on the E(Ag electrode) axis when [Ag$^+$(aq)] = 1.00 mol dm^{-3}, and lg[Ag(aq)] = 0 is 0.80 V.

FIGURE 2 The Variation of Silver Electrode Potential with Silver Ion Concentration (not to scale)

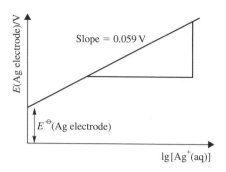

By definition (**13.1.1**), this intercept at $[Ag^+(aq)] = 1.00$ mol dm^{-3} is the standard electrode potential, $E^\ominus$ of silver.

The gradient of the graph is 0.059 V. The form of the graph is therefore,

$$E = E^\ominus + 0.059 \lg[Ag^+(aq)]$$

For the zinc electrode,

$$Zn^{2+}(aq) + 2e^- \rightleftharpoons Zn(s)$$

the gradient of the plot of E against $\lg[Zn^{2+}(aq)] = (0.059/2)$ V. In general,

$$E = E^\ominus + (0.059/z) \lg[\text{ion}]$$

where z is the charge on the metal ions.

CHECKPOINT ON ADDITIONAL TOPIC HOW FAR, HOW FAST? 3

1. (*a*) Explain the difference between electrode potential and standard electrode potential.

(*b*) The value of $E^\ominus$ for $Zn^{2+}(aq) + 2e^- \rightleftharpoons Zn(s)$ is -0.76 V Is the value of the zinc electrode potential in a 0.010 mol dm^{-3} solution of zinc sulphate (i) -0.701 V (ii) -0.731 V (iii) -0.760 V (iv) -0.790 V (v) -0.819 V?

2. Write the equation for the manganate(VII)–manganese(II) equilibrium. Deduce which will be the stronger oxidising agent, acidified manganate(VII) or neutral manganate(VII). Explain your reasoning.

3. Write the equation for the dichromate(VI)–chromium(III) equilibrium. Deduce which will be the stronger oxidising agent, acidified dichromate(VI) or neutral dichromate(VI). Explain your reasoning.

ADDITIONAL TOPIC HOW FAR, HOW FAST? 4: ADDITIONAL PRACTICE ON REACTION MECHANISMS

1. The decomposition of NO_2Cl is believed to proceed by the mechanism

Step 1 slow : $NO_2Cl \rightarrow NO_2 + Cl\cdot$

Step 2 fast : $NO_2Cl + Cl\cdot \rightarrow NO_2 + Cl_2$

If the first step is slow and the second step is fast, what is the experimental rate law?

2. **A** and **B** react to form **C**. The reaction is first order with respect to **A** and second order with respect to **B**.

(*a*) Write the rate equation.

(*b*) What is the overall order of reaction?

(*c*) What happens to the rate of reaction (i) if [**A**] is doubled while [**B**] is constant (ii) if [**B**] is doubled while [**A**] is constant (iii) if [**A**] and [**B**] are both doubled?

3. For the reaction

$$2A + B \rightarrow C + D$$

the experimental rate law is

$$\text{Rate} = k[A]^2[B]$$

A termolecular collision would fit the kinetics but is unlikely on statistical grounds. It is more usual to find that such a reaction takes place by means of a series of bimolecular processes. Suggest a reaction mechanism of this nature that will fit the rate equation.

4. For the reaction

$$2NO(g) + Br_2(g) \rightarrow 2NOBr(g)$$

the following mechanism has been suggested:

Step 1: $NO + Br_2 \rightarrow NOBr_2$

Step 2: $NOBr_2 + NO \rightarrow 2NOBr$

(a) Write the rate equation if Step 1 is slow and Step 2 is fast.

(b) Write the rate equation if Step 2 is slow and Step 1 is a rapidly established equilibrium:

$$NO + Br_2 \rightleftharpoons NOBr_2$$

5. The reaction

$$NO_2(g) + CO(g) \rightarrow CO_2(g) + NO(g)$$

is zero-order with respect to CO. A mechanism which has been proposed is:

$$2NO_2 \xrightarrow{\text{Slow}} NO_3\cdot + NO$$

$$NO_3\cdot + CO \xrightarrow{\text{Fast}} NO_2 + CO_2$$

Explain why this mechanism is consistent with zero-order kinetics with respect to CO.

6. The following results were obtained in a kinetic study of the reaction

$$(CH_3)_3Br + OH^- \rightarrow (CH_3)_3COH + Br^-$$

The initial rate of formation of $(CH_3)_3COH$ was measured in five experiments.

Experiment	Initial concentration/ mol dm^{-3}		Initial rate/mol dm^{-3} s^{-1}
	$(CH_3)_3CBr$	OH^-	
1	0.10	0.10	0.0010
2	0.20	0.10	0.0020
3	0.30	0.10	0.0030
4	0.10	0.20	0.0010
5	0.10	0.30	0.0010

(a) Write the rate law for the reaction.

(b) What does the kinetic information suggest about the reaction mechanism?

7. For the reaction

$$CH_3Br + OH^- \rightarrow CH_3OH + Br^-$$

the rate law that is obtained from kinetic measurements is

$$\text{Rate} = k[\text{CH}_3\text{Br}][\text{OH}^-]$$

This is in contrast to the rate law for the hydrolysis of $(\text{CH}_3)_3\text{CBr}$ in Question 6. Propose a mechanism that accounts for the observed kinetics.

ADDITIONAL TOPIC HOW FAR, HOW FAST? 5: ENZYMES

1. ENZYMES AS CATALYSTS

Enzymes are the substances that catalyse biological reactions. (For catalysis, see **14.10**.) They are proteins. Enzymes can bring about reactions in aqueous solution, at the pH and temperature of living organisms. They bring about reactions under these mild conditions when many of the non-enzymic reactions we are familiar with require high temperature or high pressure or high or low pH to give a good yield.

Compared with the catalysts you have met in your study of chemical reactions (**14.10**), enzymes are very **specific**; that is, an enzyme catalyses the reactions of only one substance or a very limited range of substances. The substance which an enzyme enables to react is called the **substrate**. Most enzymes act within cells, and cells contain many substances. It is important that an enzyme targets its own substrate and leaves other substances untouched.

Biological reactions are catalysed by enzymes. They are proteins, and their catalytic activity depends on their tertiary structure.

Like other proteins, enzymes have a three-dimensional configuration, the tertiary structure of the protein. The tertiary structure of the protein is dependent on pH and temperature (see Additional Topic How Far, How Fast? 6). Since the catalytic power of the enzyme depends on its tertiary structure, enzyme activity is sensitive to changes in pH and temperature.

2. SPECIFICITY

The pancreas secretes three proteolytic enzymes: trypsin, chymotrypsin and elastase. The enzymes are specific: each can catalyse the hydrolysis of peptide links between certain amino acid residues but not others. Consider the hydrolysis of this peptide:

$$-\text{N}-\underset{\underset{\text{H}}{|}}{\overset{\overset{R_1}{|}}{\text{C}}}-\underset{\underset{\text{O}}{||}}{\text{C}}-\text{N}-\underset{\underset{\text{H}}{|}}{\overset{\overset{H}{|}}{\text{C}}}-\underset{}{\overset{\overset{R_2}{|}}{\text{C}}}\overset{\overset{O}{||}}{-} + \text{H}_2\text{O} \longrightarrow$$

Enzymes are specific ...

$$-\text{N}-\underset{\underset{\text{H}}{|}}{\overset{\overset{R_1}{|}}{\text{C}}}-\text{C}\underset{\text{O}-\text{H}}{\overset{\overset{O}{\diagup\!\!\!\diagup}}{\diagdown}} + \text{H}_2\text{N}-\underset{\underset{\text{H}}{|}}{\overset{\overset{R_2}{|}}{\text{C}}}-\underset{}{\overset{\overset{O}{||}}{\text{C}}}-$$

... catalysing the reaction of one substrate only or a few related substrates.

The nature of the groups R_1 and R_2 adjacent to the peptide link to be hydrolysed decides which enzyme will catalyse the reaction. The reason for the specificity of enzymes for substrates is that the substrate must fit into the **active site** of the enzyme.

The active site: the lock and key theory

The active site of an enzyme is only a small region, perhaps 5% of the enzyme's surface.

FIGURE 3. The Substrate Binds to the Active Site of the Enzyme and Reacts

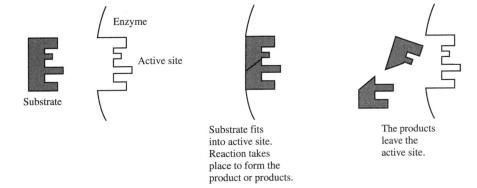

The reason for the specificity...
...is that the substrate fits the active site of the enzyme...
...as a key fits a lock.

It is a crevice in the enzyme molecule into which the substrate molecule fits. The fit between an enzyme and its substrate has been compared with that of a lock and a key. The **lock and key theory** of the active site was originated by Emil Fischer in 1894. Just as it takes the correct key to open a lock, it takes the correct enzyme to bond to the substrate and catalyse its reaction. The very precise fit is the reason why enzymes are so specific. The active site of an enzyme is composed of a **binding site** and a **catalytic site**. The bonding between enzyme and substrate may involve electrostatic attraction, hydrogen bonding and van der Waals forces.

The active site binds the substrate...
...and catalyses its reaction.

The active site of an enzyme interacts with the substrate in such a way as to weaken the bond which is to be broken. Often polar groups, such as the $-NH_2$ group and the $-CO_2H$ group, are the groups which catalyse the reaction. By attracting or repelling electrons in the substrate such groups can alter the strength of the bond which is to be broken.

The active site: the induced fit theory

There are some defects in the lock and key theory of the active site. X-ray diffraction studies show that changes in the conformation of the active site occur when it binds the substrate. In 1958 Koshland modified the theory. He suggested that the shape of the active site of the enzyme is not exactly complementary to the shape of the substrate molecule. When the substrate binds to the enzyme, however, it changes the conformation of the active site. This idea is called the **induced fit theory**. Only the substrate is capable of inducing the correct change in the active site.

FIGURE 4 A Comparison of the Lock and Key Theory and the Induced Fit Theory

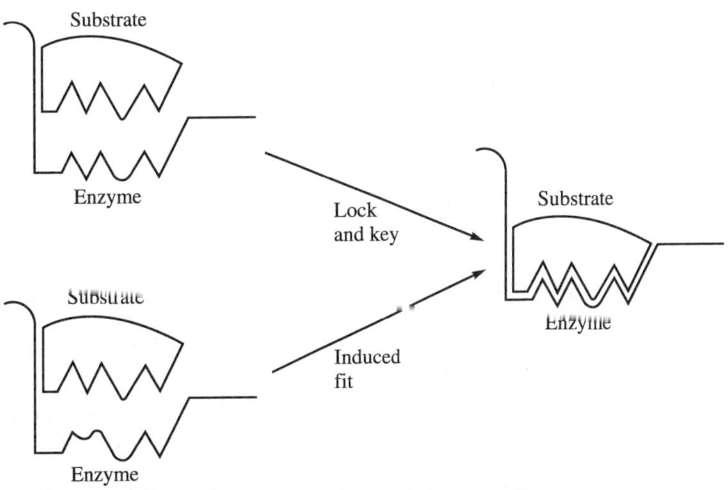

The induced fit theory of enzyme specificity has replaced the lock and key theory.

3. IMMOBILISED ENZYMES

Many industrial processes are **batch processes** which use soluble enzymes. The enzyme and substrate are mixed in a reaction vessel and left until a good yield of product has been formed. Usually the enzyme is discarded because recovery would be too costly. If the enzyme is expensive, only a small quantity may be used, and the reaction may be left to run for a long time with the possible formation of unwanted products from side reactions. The cost of the enzyme used in an industrial process can be reduced if the enzyme can be recovered and used again. Making the enzyme insoluble by attaching it to an inert support achieves this aim. The technique is called **immobilising** the enzyme. The enzyme and its support can be packed into columns and substrate can be passed over the enzyme in a **continuous process** over long periods. Alternatively, the enzyme and its support may be in the form of insoluble granules which can be used in a batch reactor and then separated by centrifugation and re-used. Immobilising the enzyme allows a higher concentration of enzyme to be used, thus reducing the reaction time, reducing the formation of unwanted side-products and reducing operational costs.

The technique of immobilising enzymes allows enzymes to be used in continuous processes.

Methods of binding the enzyme

As supports for enzymes insoluble polymers are used. Methods include the following.

Adsorption onto a resin The enzyme is adsorbed onto a resin by means of interionic attractions between amino acid residues in the enzyme and charged groups on the support material, e.g. Sephadex.

The enzyme may be adsorbed onto a resin ...

Trapping in a polymer A solution containing the enzyme and a water-soluble monomer is prepared and the monomer is allowed to polymerise. The large enzyme molecules are trapped inside a network of polymer molecules. Small substrate molecules are able to penetrate the structure and interact with the enzyme.

... or trapped inside a network of polymer molecules ...

Covalent bonding to a polymer One or more of the enzyme side-chains, e.g. $-NH_2$, $-OH$, $-CO_2H$, form covalent bonds with a polymer support material.

... or covalently bonded to a polymer.

Examples of the use of immobilised enzymes

The use of immobilised enzymes has halved the cost of production of **high-fructose syrup**. The hydrolysis of starch produces glucose. It is an advantage to convert glucose into fructose, which is a sweeter sugar. Starch is partially hydrolysed and then treated firstly with glucoamylase to give glucose and secondly with glucose isomerase to convert glucose into fructose.

Some examples of the use of immobilised enzymes.

Many strains of bacteria have become resistant to penicillin. It is possible to convert penicillin into **semi-synthetic penicillin** which will attack bacteria that are resistant to penicillin. In the industrial production, a solution of penicillin is trickled over the enzyme penicillin acylase bonded to a modified cellulose support.

See also **2.18** in *Biochemistry and Food Science* by E. N. Ramsden (Stanley Thornes, 1995).

CHECKPOINT ON ADDITIONAL TOPIC HOW FAR, HOW FAST? 5

1. Explain the following terms: catalyst, enzyme, substrate, active site.

2. What is meant by the specificity of enzyme action? Why are enzymes more specific than other catalysts?

3. Briefly explain the difference between the lock-and-key theory and the induced fit theory of the active site.

4. What is an immobilised enzyme? Explain the advantages of using immobilised enzymes over free enzymes.

ADDITIONAL TOPIC HOW FAR, HOW FAST? 6: THE BIOLOGICAL IMPORTANCE OF BUFFER SOLUTIONS

1. THE NEED TO MAINTAIN A STABLE pH

Protein molecules are held in a three-dimensional configuration by interactions between protein chains ...
... including interactions between polar groups, e.g. $-CO_2^-$ and $-NH_3^+$

Protein molecules are long chains of amino acid residues. The sequence of amino acids is the **primary structure** of the protein. The chains are coiled into helical structures or folded into pleated structures. These conformations are called the **secondary structure** of the protein. The coils or sheets are further coiled into the tertiary structure of the protein (Figure 5). You can see how forces of attraction between polar groups such as $-CO_2^-$ and $-NH_3^+$ hold the protein chain in its conformation. Should the medium become too acidic, $-CO_2^-$ groups would be converted into $-CO_2H$ groups.

$$-CO_2^-(aq) + H^+(aq) \rightleftharpoons -CO_2H(aq)$$

This would interfere with the attraction between $-CO_2^-$ groups and $-NH_3^+$ groups. Should the medium become too alkaline, $-NH_3^+$ groups would be converted into $-NH_2$ groups.

These interactions depend on the pH

$$-NH_3^+(aq) + OH^-(aq) \rightleftharpoons -NH_2(aq) + H_2O(l)$$

Again this would interfere with the attraction between $-NH_3^+$ groups and $-CO_2^-$ groups. The attractions between acidic and basic groups therefore exist only within a limited range of pH values. Outside this range, the tertiary structure of the protein is destroyed. Blood proteins would coagulate if the pH of blood were allowed to depart far from neutral pH.

Enzymes are proteins and their activity depends on their three-dimensional structure ...
... which depends on pH
Enzymes can work only within a narrow range of pH.

The reactions that take place in living things are catalysed by enzymes. Enzymes are proteins, and the ability of enzymes to act as catalysts depends on their tertiary structure. Outside the range of pH which maintains the tertiary structure of the protein, the enzyme cannot do its job. There is an optimum pH for each enzyme, and the enzyme will only function over a range of 2 or 3 pH units. Most cells and tissues require a pH value close to 7.

2. HOW CELL CONTENTS ARE BUFFERED

Mechanisms exist to keep the cell pH as constant as possible. This is achieved by buffers (**12.7.12**). Many tissues contain hydrogenphosphate ions and, since phosphoric acid is a weak acid, the equilibrium is set up:

$$H_2PO_4^-(aq) \rightleftharpoons H^+(aq) + HPO_4^{2-}(aq)$$

Cell contents are buffered to provide a suitable pH ...
... by phosphates ...

When the hydrogen ion concentration increases, the equilibrium is displaced from right to left, thus removing hydrogen ions from solution. If the hydrogen ion concentration decreases, $H_2PO_4^-$ ions dissociate to release H^+ ions. Thus the buffer tends to maintain a constant hydrogen ion concentration.

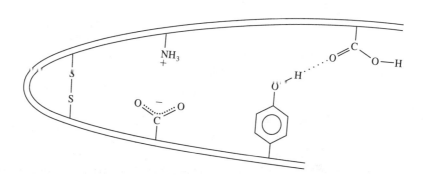

FIGURE 5 Some Interactions between Groups in a Protein Molecule

Blood contains carbonate ions and hydrogencarbonate ions in equilibrium:

$$HCO_3^-(aq) \rightleftharpoons H^+(aq) + CO_3^{2-}(aq)$$

The system acts as a buffer, absorbing small amounts of hydrogen ion by moving from right to left and releasing small amounts of hydrogen ion by moving from left to right.

... by hydrogencarbonates ...
If hydroxide ions accumulate in the tissue, hydrogencarbonate ions can absorb them by forming carbonate ions and water:

$$HCO_3^-(aq) + OH^-(aq) \rightleftharpoons CO_3^{2-}(aq) + H_2O(l)$$

Amino acids have the formula $H_2NCHRCO_2H$, where R can be a number of different groups. They can ionise as weak acids and as weak bases. Both the weakly acidic carboxyl group and the weakly basic amino group can act as buffers.

$$H_2NCHRCO_2H(aq) + H_2O(l) \rightleftharpoons H_3N^+CHRCO_2H(aq) + OH^-(aq)$$

$$H_2NCHRCO_2H(aq) + H_2O(l) \rightleftharpoons H_2NCHRCO_2^-(aq) + H_3O^+(aq)$$

At a certain pH the ionisation as an acid and the ionisation as a base are balanced.

$$H_2NCHRCO_2H(aq) \rightleftharpoons H_3N^+CHRCO_2^-(aq)$$

... by amino acids, peptides and proteins.
This pH is called the **isoelectric point**, and the ion is called a **zwitterion** (33.17.1). Peptides and proteins are condensation polymers of amino acids (33.17.2) and share this buffering action. Amino acids, peptides and proteins are important buffering agents in living tissues.

CHECKPOINT ON ADDITIONAL TOPIC HOW FAR, HOW FAST? 6

1. Write the equilibrium for the ionisation of the weak acid RCO_2H. Explain how a solution of the acid can absorb small amounts of (*a*) hydrogen ions (*b*) hydroxide ions with only a very small change in pH.

2. Write the equilibrium for the ionisation of the weak base RNH_2. Explain how a solution of the base can absorb small amounts of (*a*) hydrogen ions (*b*) hydroxide ions with only a very small change in pH.

3. Glycine, $H_2NCH_2CO_2H$, is the simplest amino acid.

(*a*) Write an equation to show how glycine ionises (i) as an acid (ii) as a base.

(*b*) Write an equation to show how glycine can combine with small amounts of hydrogen ions without a significant change in pH.

(*c*) Write an equation to show how glycine can combine with small amounts of hydroxide ions without a significant change in pH.

4. Briefly explain why a fairly constant pH is important for living tissues.

EXAMINATION QUESTIONS ON HOW FAR, HOW FAST?

1. Hydrogen and iodine are mixed together in a sealed container at constant temperature and the mixture allowed to reach equilibrium.

$$H_2(g) + I_2(g) \rightleftharpoons 2HI(g) \quad \Delta H = +26.5 \text{ kJ mol}^{-1}$$

(*a*) (i) At a particular temperature the equilibrium constant, K_c, is 55. Write an expression for K_c.
(2 marks)
(ii) At this temperature the concentration of $H_2(g)$ is 5.10×10^{-3} mol dm^{-3} and that of HI(g) is 3.67×10^{-3} mol dm^{-3}. Calculate the equilibrium concentration of $I_2(g)$.
(2 marks)

(*b*) What would be the effect on the value of K_c of:
(i) increasing the temperature of the equilibrium mixture?
(3 marks)
(ii) increasing the concentration of $H_2(g)$?
(3 marks)
(iii) increasing the pressure of the mixture?
(2 marks)
(Total 12 marks)
[C, '93 How Far, How Fast?]

2. The rate of a chemical reaction depends on many variables.

(*a*) Explain how each of the following, when varied individually, could influence the rate of a given reaction:
(i) concentration;
(ii) temperature;
(iii) a catalyst.
You should refer to activation energy and the Boltzmann distribution, where relevant.
(12 marks)

(*b*) (i) Explain what you understand by the terms *rate determining step* and *order of reaction*.
(ii) The reaction,

$$2A + 2B \rightarrow C + D$$

is known to proceed via a two step mechanism. Its rate equation was found to be

rate = $k[A][B]^2$

Suggest a possible mechanism.
(7 marks)

(c) Iodine reacts with propanone, $(CH_3)_2CO$, in the presence of an acid catalyst, $H^+(aq)$, in accordance with the equation

$$I_2(aq) + (CH_3)_2CO(aq) \xrightarrow{H^+(aq)} CH_3COCH_2I(aq) + H^+(aq) + I^-(aq)$$

The rate equation for this reaction is

rate = $k[(CH_3)_2CO(aq)][H^+(aq)]$

Explain how you would deduce the rate equation experimentally, indicating how you would
(i) control and regulate the experimental variables,
(ii) monitor the rate of the reaction.
(6 marks)
[C, '95]

3. Beer was brewed by the ancient Egyptians and is thought to have been among the rations of the builders of the Pyramids. The composition of a beer is given in the table.

Constituent	Mass per dm³/g
ethanol, C_2H_5OH	30
glucose, $C_6H_{12}O_6$	20
protein	3

Ethanol is a food as well as a drug and it is a more concentrated energy source than carbohydrate.

(a) Write a balanced equation, including state symbols, for the complete combustion of ethanol.
(2 marks)

(b) (i) Define the term *standard enthalpy change of combustion*.
(3 marks)

The standard enthalpy change of combustion of ethanol and of glucose are given as -1370 kJ mol^{-1} and -3000 kJ mol^{-1}, respectively.
(ii) Calculate the energy per gram of ethanol and of glucose.
(4 marks)
(iii) Calculate the total energy released by the combustion of the ethanol and glucose contained in 1 dm³ of the beer, as detailed in the table.
(2 marks)

(c) Beer is generally made from malt, hops, yeast and water, but sometimes sugar is added to increase the fermentable material. Generally, 'invert sugar' (sucrose which has been hydrolysed to glucose and fructose) is used. Sucrose decomposes into the fructose and glucose by a first-order reaction with a half-life of 3.00 hours at 25 °C.
(i) Sketch a graph of the percentage of sucrose against time, during the hydrolysis.
(2 marks)
(ii) Use your graph to determine the percentage of sucrose that would be left after 8 hours.
(1 mark)
[C, '95]

4. The rate of reaction between peroxodisulphate ions, $S_2O_8^{2-}$, and iodide ions, I^-, in aqueous solution

$$S_2O_8^{2-}(aq) + 2I^-(aq) \rightarrow 2SO_4^{2-}(aq) + I_2(aq)$$

may be studied by measuring the amount of iodine formed at fixed intervals of time. The data below were obtained from three separate experiments **A**, **B** and **C** carried out at constant temperature.

	Initial conc. of $S_2O_8^{2-}$ (aq) /mol dm^{-3}	Initial conc. of I^- (aq)/ mol dm^{-3}	Initial rate/ mol dm^{-3} s^{-1}
A	0.0100	0.200	4.10×10^{-6}
B	0.0200	0.200	8.20×10^{-6}
C	0.0200	0.400	1.64×10^{-5}

(a) (i) Show how these data can be used to deduce the order with respect to each reactant and the overall rate equation for the reaction between the $S_2O_8^{2-}$(aq) ions and the I^-(aq) ions. Explain all the terms in the rate equation and calculate a value for the rate constant.
(ii) Suggest a suitable method for determining the amount of iodine produced during the reaction and explain why the above kinetic investigation was carried out at constant temperature.
(iii) For a mixture of $S_2O_8^{2-}$(aq) ions and I^-(aq) ions, sketch a graph to show how the concentration of iodine, I_2, varies with time. Explain how the initial rate of reaction could be measured.
(16 marks)

(b) Radioactive decay shows first-order reaction kinetics. Archaeologists can determine the age of organic matter by measuring the proportion of radioactive carbon-14 present. Write the rate equation for the decay of ^{14}C and estimate the age of a piece of wood found to contain $\frac{1}{8}$th as much ^{14}C as living material.
(Assume that carbon-14 has a half-life of 5600 years.)
(4 marks)

(c) Catalysts can influence the rate of a chemical reaction. Explain:
(i) why transition metals can often function as catalysts;
(ii) the difference between a heterogeneous catalyst and a homogeneous catalyst, giving a suitable example of each.
(5 marks)
[C, '93]

5. (a) (i) Explain what is meant by a Bronsted-Lowry acid. Explain the differences in acidic behaviour of hydrochloric acid and ethanoic acid, CH_3CO_2H.
(ii) The concentration of an acid can be monitored by titration with an alkali such as aqueous sodium hydroxide.
Sketch a curve in each case to show the change in pH when 0.10 mol dm^{-3} aqueous sodium hydroxide is added, until in excess, separately to 25 cm³ portions of 0.10 mol dm^{-3} aqueous hydrochloric acid and of 0.10 mol dm^{-3} aqueous ethanoic acid.
(iii) Methyl orange has a pH range of approximately 3.0 to 5.0. State and explain whether or not it would be a suitable indicator for each titration.
(12 marks)

(b) A common method for commercially peeling potatoes is to soak them for a short time in aqueous sodium hydroxide at about 75 °C and then to spray off the peel once the potatoes are removed from the solution. In order to be effective the aqueous sodium hydroxide must have a minimum concentration of 2.0 mol dm^{-3}. A 10.0 cm^3 sample of the alkali solution is titrated at regular intervals with a 0.20 mol dm^{-3} solution of sulphuric acid. Calculate the volume of sulphuric acid that would indicate that the aqueous sodium hydroxide solution should be replaced.

(5 marks)

(c) Indicators are often weak acids and can be represented by the formula HIn such that

$$HIn(aq) \rightleftharpoons H^+(aq) + In^-(aq)$$

The indicator equilibrium constant, K_{In}, is equivalent to K_a for an acidic indicator. Calculate the pH of a 0.10 mol dm^{-3} aqueous bromophenol. Bromophenol has a pK_{In} (i.e. pK_a) value of 4.0.

(8 marks)
[C, '95]

6. (a) What is meant by a *dynamic equilibrium*?

(3 marks)

(b) In the Contact process for the preparation of sulphuric acid (H$_2$SO$_4$), sulphur dioxide, SO$_2$, is converted to sulphur trioxide, SO$_3$, in accordance with:

$$2SO_2(g) + O_2(g) \rightleftharpoons 2SO_3(g) \quad \Delta H = -94.6 \text{ kJ mol}^{-1}$$

(i) State *Le Chatelier's Principle* and use the principle to predict the optimum theoretical temperature and pressure to ensure the maximum possible yield of sulphur trioxide, SO$_3$.
(ii) Typical industrial conditions employed are a temperature of around 450 °C at atmospheric pressure in the presence of vanadium(V) oxide, V$_2$O$_5$, as a catalyst. Compare these conditions with the optimum theoretical conditions and justify their use.
(iii) At 450 °C the partial pressures of the gases in the equilibrium mixture are:

p_{SO_2}, 0.090 atm; p_{SO_3}, 4.5 atm; p_{O_2}, 0.083 atm.

Write an expression for the equilibrium constant, K_p, and calculate its value under these conditions.

(14 marks)

(c) Gaseous sulphur dioxide is a major pollutant. A possible sequence for the conversion of SO$_2$(g) into SO$_3$(g) is:

$$NO(g) + \tfrac{1}{2}O_2(g) \rightarrow NO_2(g)$$
$$NO_2(g) + SO_2(g) \rightarrow NO(g) + SO_3(g)$$

(i) State the role played by the NO(g) in this sequence.
(ii) Write an equation to show how the SO$_3$(g) is converted into sulphuric acid and state **two** possible effects of the acid rain produced on the environment.

(4 marks)

(d) The combustion of petrol in cars causes numerous forms of pollution of which NO(g) and NO$_2$(g) are two. More and more cars are being fitted with catalytic converters which 'convert' these oxides back to their elemental form. The catalytic converter contains a fine-meshed aluminium alloy coated in a platinum-rhodium mixture.
(i) Write equations to show the decomposition of the nitrogen oxides into their elements.
(ii) Explain why the catalytic converter contains a fine mesh.
(iii) Discuss the limitations on the type of petrol that can be used in a car fitted with a catalytic converter.

(4 marks)
[C, '93]

7. Ammonia is manufactured from nitrogen and hydrogen by the Haber process. Nitrogen and hydrogen are obtained, by a multi-stage process, from methane, steam and air. The reaction between nitrogen and hydrogen is exothermic and incomplete. The ammonia formed is removed from the mixture and the unreacted gases are recycled. The process normally operates at about 500 °C and 200 atm.

(a) (i) Write an equation for the manufacture of ammonia by the Haber process.

(1 mark)

(ii) Why is the reaction carried out at a high pressure?

(1 mark)

(iii) State **one** advantage and **one** disadvantage of carrying out the reaction at high temperature.

(2 marks)

(b) (i) Write an expression for the equilibrium constant, K_p, for the ammonia synthesis.

(1 mark)

(ii) Calculate the value of K_p given the following partial pressures in the table which apply at 500 °C and a total pressure of 200 atm.

Gas	Partial pressure/atm
nitrogen	75
hydrogen	35
ammonia	90

(2 marks)

(iii) State the units of K_p in (b) (ii).

(1 mark)

(iv) Iron is used as a catalyst in the ammonia manufacture. State its effect on the K_p value you have calculated in (b) (ii).

(1 mark)

(c) Urea, CO(NH$_2$)$_2$, is a naturally occurring substance which can be hydrolysed with water to form ammonia according to the following equation.

$$H_2O(l) + CO(NH_2)_2(aq) \rightarrow CO_2(aq) + 2NH_3(aq)$$

The above reaction only proceeds at a detectable rate in the presence of the enzyme urease.
Explain why enzymes, when they function as biological catalysts, have such a specific activity.

(2 marks)

(d) The standard enthalpy changes of formation of water, urea, carbon dioxide and ammonia are given in the table.
Use these data to calculate the standard enthalpy change for the hydrolysis reaction.

Compound	$\Delta H_f^\ominus$/ kJ mol^{-1}
H$_2$O(l)	−287.0
CO(NH$_2$)$_2$(aq)	−320.5
CO$_2$(aq)	−414.5
NH$_3$(aq)	−81.0

(2 marks)
[C, '95]

8. The reaction of ethanoic anhydride, $(CH_3CO)_2O$, with ethanol, C_2H_5OH, can be represented by the equation

$$(CH_3CO)_2O + C_2H_5OH \rightarrow CH_3CO_2C_2H_5 + CH_3CO_2H$$

The table shows the initial concentrations of the two reactants and the initial rate of the reaction. All measurements were carried out at the same temperature and the reaction was carried out using hexane as the solvent.

Experiment	$[(CH_3CO)_2O]/$ mol dm^{-3}	$[C_2H_5OH]/$ mol dm^{-3}	Initial rate/ mol dm^{-3} s^{-1}
1	0.200	0.200	3.30×10^{-4}
2	0.400	0.400	1.32×10^{-3}
3	0.800	0.400	2.64×10^{-3}

(*a*) State and explain the order of reaction with respect to
(i) ethanoic anhydride
(2 marks)

(ii) ethanol
(3 marks)

(*b*) (i) Write an expression for the overall rate equation.
(1 mark)

(ii) What is the overall order of reaction?
(1 mark)

(iii) Calculate the value for the rate constant, *k*.
(2 marks)

(*c*) When the same experiments were repeated using ethanol as the solvent, in place of hexane, the rate equation obtained was

$$\text{rate} = k[(CH_3CO)_2O]$$

Suggest an explanation for any difference in the rate gains in parts (*b*) (i) and (*c*).
(2 marks)
[C, '95]

4827: BIOCHEMISTRY

Note that the references in this module are to *Biochemistry and Food Science* by E. N. Ramsden (Stanley Thornes, 1995).

1. PROTEINS

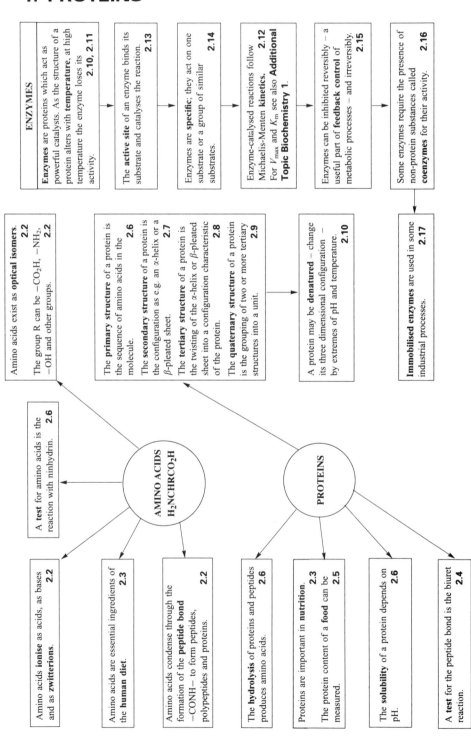

2. CARBOHYDRATES

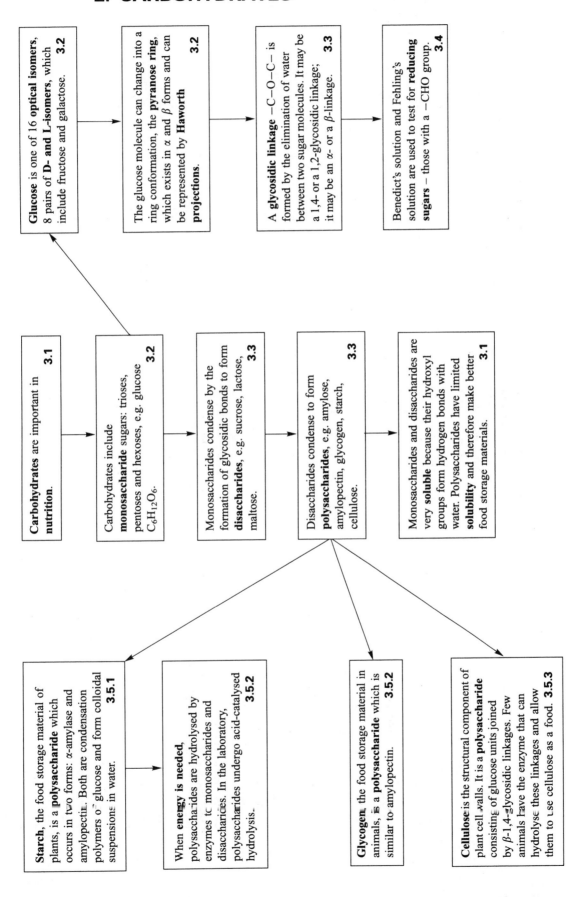

3. LIPIDS AND MEMBRANE STRUCTURE

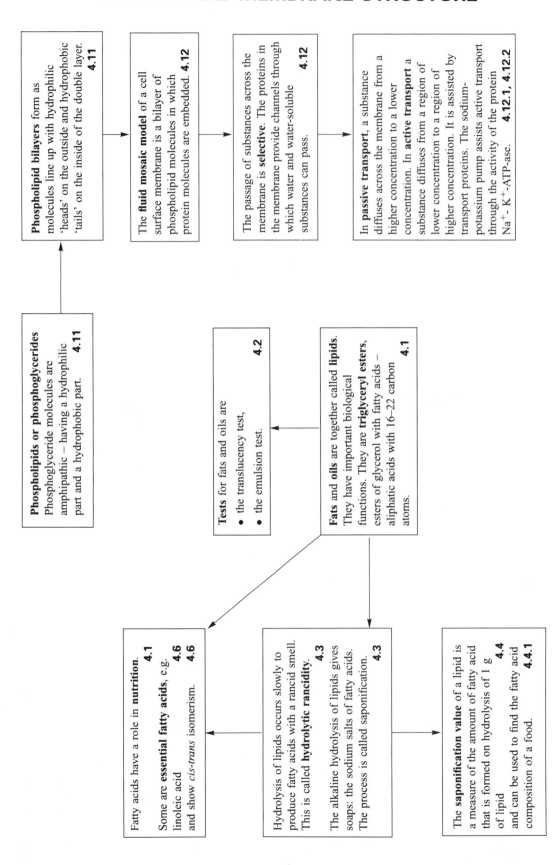

Phospholipids or phosphoglycerides Phosphoglyceride molecules are amphipathic – having a hydrophilic part and a hydrophobic part. **4.11**

Phospholipid bilayers form as molecules line up with hydrophilic 'heads' on the outside and hydrophobic 'tails' on the inside of the double layer. **4.11**

The **fluid mosaic model** of a cell surface membrane is a bilayer of phospholipid molecules in which protein molecules are embedded. **4.12**

The passage of substances across the membrane is **selective**. The proteins in the membrane provide channels through which water and water-soluble substances can pass. **4.12**

In **passive transport**, a substance diffuses across the membrane from a higher concentration to a lower concentration. In **active transport** a substance diffuses from a region of lower concentration to a region of higher concentration. It is assisted by transport proteins. The sodium-potassium pump assists active transport through the activity of the protein Na^+-K^+-ATP-ase. **4.12.1, 4.12.2**

Tests for fats and oils are
- the translucency test,
- the emulsion test. **4.2**

Fats and **oils** are together called **lipids**. They have important biological functions. They are **triglyceryl esters**, esters of glycerol with fatty acids – aliphatic acids with 16–22 carbon atoms. **4.1**

Fatty acids have a role in **nutrition**. **4.1**
Some are **essential fatty acids**, e.g. linoleic acid **4.6**
and show *cis-trans* isomerism. **4.6**

Hydrolysis of lipids occurs slowly to produce fatty acids with a rancid smell. This is called **hydrolytic rancidity**. **4.3**
The alkaline hydrolysis of lipids gives soaps: the sodium salts of fatty acids. The process is called saponification. **4.3**

The **saponification value** of a lipid is a measure of the amount of fatty acid that is formed on hydrolysis of 1 g of lipid **4.4**
and can be used to find the fatty acid composition of a food. **4.4.1**

4. NUCLEIC ACIDS

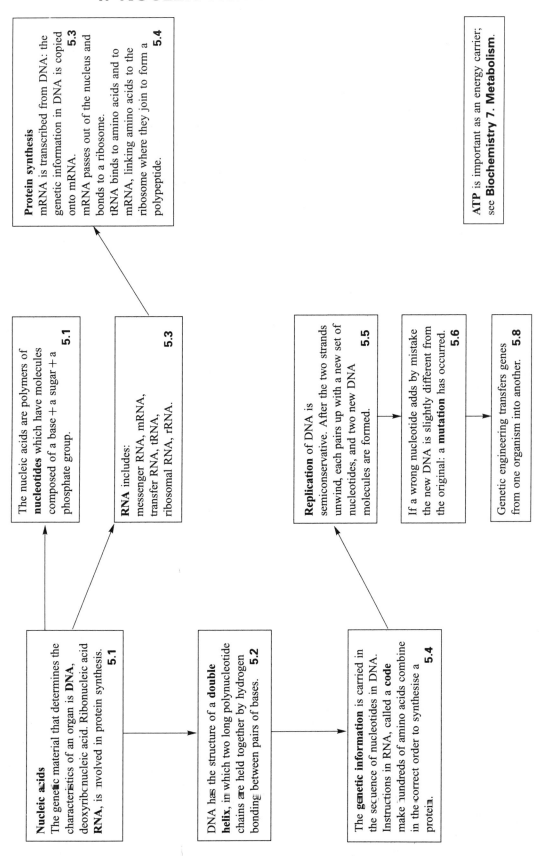

Nucleic acids
The genetic material that determines the characteristics of an organ is **DNA**, deoxyribonucleic acid. Ribonucleic acid **RNA**, is involved in protein synthesis. **5.1**

The nucleic acids are polymers of **nucleotides** which have molecules composed of a base + a sugar + a phosphate group. **5.1**

DNA has the structure of a **double helix**, in which two long polynucleotide chains are held together by hydrogen bonding between pairs of bases. **5.2**

The **genetic information** is carried in the sequence of nucleotides in DNA. Instructions in RNA, called a **code** make hundreds of amino acids combine in the correct order to synthesise a protein. **5.4**

RNA includes: messenger RNA, mRNA, transfer RNA, tRNA, ribosomal RNA, rRNA. **5.3**

Protein synthesis
mRNA is transcribed from DNA: the genetic information in DNA is copied onto mRNA. **5.3**
mRNA passes out of the nucleus and bonds to a ribosome.
tRNA binds to amino acids and to mRNA, linking amino acids to the ribosome where they join to form a polypeptide. **5.4**

Replication of DNA is semiconservative. After the two strands unwind, each pairs up with a new set of nucleotides, and two new DNA molecules are formed. **5.5**

If a wrong nucleotide adds by mistake the new DNA is slightly different from the original: a **mutation** has occurred. **5.6**

Genetic engineering transfers genes from one organism into another. **5.8**

ATP is important as an energy carrier; see **Biochemistry 7. Metabolism**.

5. VITAMINS AND 6. MINERALS

Vitamins include a wide variety of compounds. **6.1** Some are water-soluble; others are fat-soluble. **6.2** A diet which is deficient in a vitamin results in disease. **6.1, 6.2** Some foods have vitamins added to them. **6.3**	**Minerals** which are important in the diet include • calcium salts, **7.2** • magnesium salts, **7.8** • iron salts, **7.3 and Additional Topic Biochemistry 2** • phosphates, **7.4** • sodium and potassium salts. **7.5** Minerals are added to some foods. Some minerals are removed by complexing agents. **7.2, 7.3** Metal ions may contaminate drinking water. **7.9**

7. METABOLISM

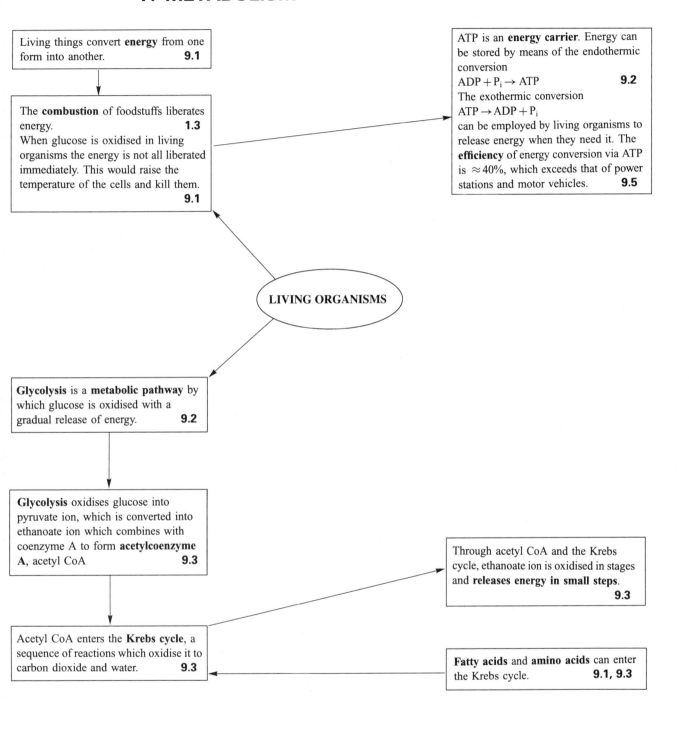

MODULE 4827: BIOCHEMISTRY

Note that the references in this module are to *Biochemistry and Food Science* by E. N. Ramsden (Stanley Thornes, 1995).

1. Proteins

(a) the general formula and properties of α-amino acids $RCH(NH_2)CO_2H$	2.2
(b) essential amino acids	2.3
(c) zwitterions	2.2
(d) peptide linkage between α-amino acids to form polypeptides and proteins; the primary structure of proteins	2.2, 2.3, 2.6
(e)(g) the hydrolysis of proteins and peptides and the separation of the products	2.6.1–3
(f) the biuret test	2.4
(h) the secondary structure of proteins: α-helix and β-pleated sheet	2.7
(i) the tertiary protein structure	2.8
(j) the quaternary structure of proteins e.g. haemoglobin	2.9
(k)(l) denaturation of proteins	2.10
(m) enzymes as catalysts	2.11, 2.14
(n) the relationship between enzyme and substrate concentration; graphical determination of V_{max} and K_m, the Michaelis constant	2.12
(o) the significance of K_m for hexokinase and glucokinase	See **Additional Topic Biochemistry 1**
(p) the active site of an enzyme	2.13
(q) competitive and non-competitive inhibition of enzymes	2.15
(r) the importance of coenzymes and cofactors	2.16
(s) competitive and non-competitive inhibition in metabolism	2.15
(t) advantages of immobilising enzymes	2.18
(u) industrial and commercial uses of enzymes	2.17

2. Carbohydrates

(a) the α- and β-pyranose ring structures of glucose; examples of a triose, a pentose and a hexose	3.2, 3.3
(b) why some sugars are reducing agents whereas others are not	3.4
(c) use of Benedict's solution	3.4
(d) D and L sugars; α-D- and β-D-glucose	3.2
(e) describe the structure of sucrose	3.3
(f) the effect of hydrogen bonding on the solubilities of carbohydrates and for hydrogen bonding	3.1 See **A-level Chemistry** 30.2.2
(g) the structures, properties and importance of cellulose, starch and glycogen	3.5

3. Lipids and membrane structure

(a) the structure and function of triglyceryl esters	4.1–9
(b) how fatty acids are obtained from triglycerides	4.3
(c) linoleic acid – an essential fatty acid	4.6
(d) the role of fatty acids in nutrition	4.1
(e) phosphoglycerides in micelles and bimolecular layers	4.11, 4.14
(f) the fluid mosaic structure of membranes	4.12
(g) active transport e.g. the Na^+/K^+ pump	4.12.2

4. Nucleic acids

(a) the structure of nucleotides and nucleic acids	5.1
(b) the chemical and physical differences between DNA and RNA	5.2, 5.3
(c) the role of DNA as the repository of genetic information	5.4–6
(d) the role of mRNA in protein synthesis	5.6
(e) the importance of ATP in metabolic activity	9.2, 9.5
(f) genetic engineering	5.8

5. Vitamins

(a) vitamins	6.1
(b) oil-soluble and water-soluble vitamins	6.1
(c) vitamin C	6.2
(d)(e) the effect of a deficiency in vitamins A, B_6, C and D and sources of these vitamins	6.1
(f) vitamins as food additives	6.3

6. Metals		(d) iron; anaemia	7.3
(a) metals found in the human body	7.1	(e) metal ions in drinking water	7.9
(b) calcium	7.2		
(c) copper or iron in oxidase	Additional Topic Biochemistry 2		

7. Metabolism		(e) the efficiency of energy production in the Embden–Meyerhof pathway and the Krebs cycle	9.5
(a) energy transfer between molecules and possible uses in an organism	9.1	(f) acetyl coenzyme A in energy metabolism	9.1, 9.3
(b) glycolysis	9.2	(g) the role of lipids in energy metabolism	4.1, 4.9, 9.1
(c) the anaerobic production of energy in the Krebs cycle	9.3		
(d) ATP as an 'energy store'	9.5		

ADDITIONAL TOPIC BIOCHEMISTRY 1: HEXOKINASES AND GLUCOKINASE

The reaction

$$\text{Hexose} + \text{ATP} \rightarrow \text{Hexose-6-phosphate} + \text{ADP} + \text{H}^+(\text{aq})$$

is catalysed by a group of enzymes called **hexokinases**. A **kinase** is an enzyme which catalyses a phosphorylation of a sugar. The reaction is the first step in glycolysis (**9.2** and **Figure 9.2A**). One of the hexokinases is called **glucokinase** because it is specific for the phosphorylation of glucose:

$$\text{ATP} + \text{D-Glucose} \rightarrow \text{ADP} + \text{D-Glucose-6-phosphate}$$

Some values of K_m for hexokinases are:

An enzyme which is highly specific for a substrate has a low value of K_m for that substrate.

Hexokinases in yeast and mammalian brain: K_m for glucose = K_m for fructose = 1.5×10^{-4} mol dm^{-3}

Hexokinase in liver: K_m for glucose = 1.8×10^{-2} mol dm^{-3}

Glucokinase in liver: K_m for glucose = 1.0×10^{-5} mol dm^{-3}.

Glucokinase is a hexokinase which is highly specific for glucose.

A low value of K_m (**2.12**) indicates that the relationship between the enzyme and its substrate is highly specific; a low concentration of substrate can occupy all the available active sites on the enzyme and 'saturate' the enzyme.

ADDITIONAL TOPIC BIOCHEMISTRY 2: IRON IN HAEMOGLOBIN AND CYTOCHROME

For haemoglobin see **24.13.8** and **Figure 24.16** in *A-level Chemistry* by E. N. Ramsden (Stanley Thornes).

Iron is an important component of haemoglobin and of cytochromes.

A **cytochrome** is an electron-transporting protein that contains a haem group, as does haemoglobin. The haem group in cytochromes contains an iron ion which alternates between Fe(III) and Fe(II). As it does so it is able to accept electrons from one substance and donate them to another, functioning as part of the **electron transport chain**. This is the last step in the oxidation of glucose in cellular respiration.

4843: COMPLEMENTARY MODULE: FOOD TECHNOLOGY

Note that the references in this module are to *Biochemistry and Food Science* by E. N. Ramsden (Stanley Thornes, 1995).

1. COMPOSITION AND FUNCTION OF FOOD

The **nutrients in foods** are:	
proteins,	2.1, 2.3
carbohydrates,	3.1
fats,	4.1
vitamins,	6.1
minerals,	7.1–8
water.	8.1, 8.3

Amino acids $H_2NCHRCO_2H$
Amino acids are important ingredients of the **human diet**. Some are **essential amino acids**.   2.3

The **primary structure** of a protein is the sequence of amino acids in the molecule.   2.6
Proteins may be globular or fibrous.   2.1, 2.8
They have a complicated **three-dimensional structure**.   2.7–9
A protein may be **denatured** – have its three-dimensional configuration changed – by extremes of pH and temperature.   2.10

PROTEINS

Amino acids condense through the formation of the **peptide bond** to form peptides and proteins.   2.2

Proteins are vital in **nutrition**.   2.3
The protein content of a **food** can be measured.   2.5

The **hydrolysis** of proteins and peptides produces amino acids.   2.6

Test for the peptide bond: the biuret reaction.   2.4

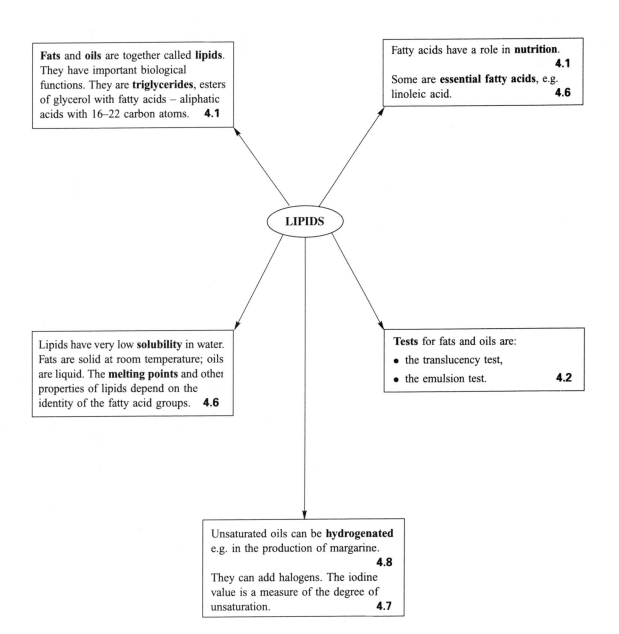

VITAMINS
Vitamins are essential components of the diet. The required daily intake is small. **6.1, 6.2**

MINERALS
Minerals in varying quantities are essential components of the diet. **7.2–6**

Toxic metals, e.g. Hg, Pb, can enter the food chain. **7.9**

WATER
Water plays an important role in the body as the medium in which all biochemical reactions take place. **8.1, 8.2**

Some foods form **colloidal dispersions** and **emulsions** in water. **8.3**

CALORIFIC VALUES
The calorific values of foods can be measured. **1.3, 3.1, 4.9**

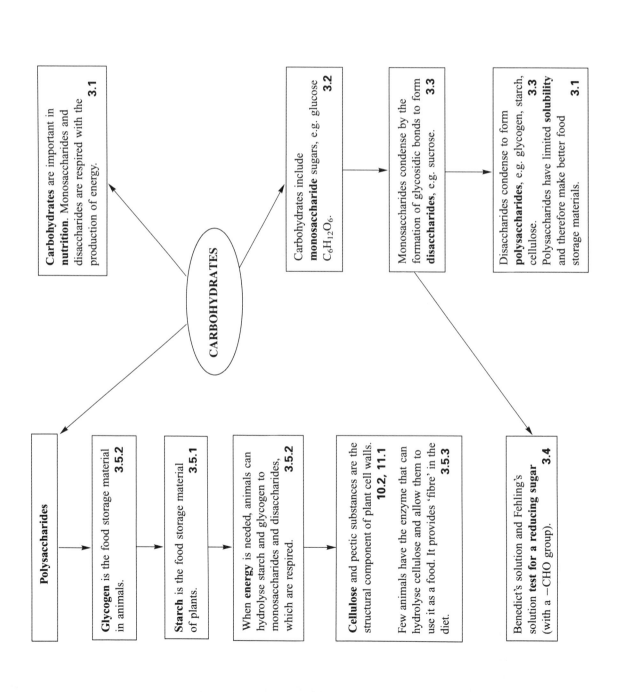

CARBOHYDRATES

Carbohydrates are important in **nutrition**. Monosaccharides and disaccharides are respired with the production of energy. **3.1**

Carbohydrates include **monosaccharide** sugars, e.g. glucose $C_6H_{12}O_6$. **3.2**

Monosaccharides condense by the formation of glycosidic bonds to form **disaccharides**, e.g. sucrose. **3.3**

Disaccharides condense to form **polysaccharides**, e.g. glycogen, starch, cellulose. **3.3**
Polysaccharides have limited **solubility** and therefore make better food storage materials. **3.1**

Polysaccharides

Glycogen is the food storage material in animals. **3.5.2**

Starch is the food storage material of plants. **3.5.1**

When **energy** is needed, animals can hydrolyse starch and glycogen to monosaccharides and disaccharides, which are respired. **3.5.2**

Cellulose and pectic substances are the structural component of plant cell walls. **10.2, 11.1**

Few animals have the enzyme that can hydrolyse cellulose and allow them to use it as a food. It provides 'fibre' in the diet. **3.5.3**

Benedict's solution and Fehling's solution **test for a reducing sugar** (with a –CHO group). **3.4**

COMPLEMENTARY MODULE: FOOD TECHNOLOGY

2. FOOD ADDITIVES

Approved **food additives** have E numbers. **13.5**

↓

Food additives are used as preservatives, **12.4.7, 13.1** colourings, flavourings, flavour-enhancers, emulsifiers, stabilisers, preservatives and thickening agents. **13.1–4**

↓

Economics
There is a need for **processed food** in modern society.
- Food processing adds to the price of a food.
- Food processing enables foods to be preserved, avoiding both wastage and food-poisoning.
- The cost of **transporting** foods from producer to processor to retailer to consumer adds to its cost. **12.5**

3. FOOD QUALITY

The **quality of beef** depends on a number of factors: breed, gender, feeding, **Additional Topic Food 1** and growth promoters. **9.8.2**

↓

The **texture of meat** is related to the structure of muscle tissue. Red and white meat differ in structure. **10.3, 11.4**

Various methods are employed to maximise **crop production**, **Additional Topic Food 2** including plant growth hormones **9.8.4**

↓

The quality of **fruits and vegetables** depends on the varieties selected and on environmental factors. **Additional Topic Food 3**

↓

Fruit may be picked unripe, **stored** in a controlled environment and **ripened** artificially. **Additional Topic Food 4 and 9.8.4**

4. FOOD STORAGE

Food spoilage takes place
- by autolysis due to the presence of enzymes in food, **12.1, 12.2**
- by microbial spoilage due to attack by moulds, yeasts and bacteria. **12.1, 12.3**

Fats suffer **hydrolytic rancidity**, **oxidative rancidity**, **4.3, 4.5** and **autoxidation** **12.2, 12.4** which is retarded by anti-oxidants, e.g. sulphites.

Enzymic browning of fruits and vegetables can be retarded by blanching. **12.2, 12.3.2**

Non-enzymic browning reactions can give foods an appetising smell **11.3.4 and Additional Topic Food 6** or they can be retarded by sulphite ions. **12.4.7**

Meat contains the protein **myoglobin** which is purple. It combines with oxygen to form **oxymyoglobin**, which is bright red. To allow oxygen to reach meat and keep it an attractive red colour, meat is packaged in 'breathing film'. **11.4**

Food poisoning may be due to bacterial toxins, e.g. from *Clostridium botulinum* or bacterial infestations, e.g. *Salmonella*. **12.3.1, 12.4.3, 12.5, Chapter 13**

5. FOOD PROCESSING

Cooking meat
Most of the meat that we eat is **skeletal muscle**. When meat is cooked, collagen and elastin are converted into gelatin, which is readily digestible. **10.3** Proteins are denatured by cooking. **2.10**

On **cooking** meat turns brown as iron(II) ions in myoglobin are oxidised to iron(III). **11.4**

Cooking vegetables
Boiling vegetables partially hydrolyses pectins, making the vegetables softer. **10.2, 11.1**

The **firmness** of boiled vegetables increases when calcium ion is added to the water. **10.6.1–2, 11.1** When vegetables are boiled in water, there is a loss of water-soluble **vitamins** **6.2** and **minerals**. **7.1–3, 7.5, 7.6**

Food processing methods include pickling, cooking, chilling, freezing, dehydration, vacuum-packing, curing, irradiation, sterilisation and pasteurisation. **12.4**

Micro-organisms are used in the food industry, e.g.
- in baking, **11.3**
- in cheese-making, **12.3, 12.4.9, 12.4.10**
- in vinegar making. **12.3**

Baking bread
Wheat grains are milled to make flour. **11.2.1**
Flour and water and yeast are mixed to form dough; Physical and chemical changes occur during **baking**. **Enzymes** in the yeast ferment sugars in the dough with the formation of carbon dioxide which makes the dough 'rise'. **11.3.3** On the surface of the bread **browning** occurs due to
- caramelisation of sugar,
- a Maillard reaction. **11.3.4**

Sugars are used in the food industry as flavourings, in fruit preservation, in leavening and in caramelisation.
10.4, 10.6.2, 11.3.2, 11.3.4

COMPLEMENTARY MODULE 4843: FOOD TECHNOLOGY

Note that the references in this module are to *Biochemistry and Food Science* by E. N. Ramsden (Stanley Thornes Publishers, 1995).

1. Composition and function of food

(a)(b) the nutrients in foods and their functions:	
proteins,	2.1, 2.3
carbohydrates,	3.1
fats,	4.1
vitamins,	6.1
minerals,	7.1–8
and water	8.1, 8.3
(c) primary, secondary, tertiary, quaternary structure of proteins; globular and fibrous proteins	2.1, 2.7, 2.8
(d) total protein content and essential amino acids	2.2, 2.3 including **Table 2.2A**, 2.4, 2.5
(e) the structure of an amino acid,	2.2
the folding of a protein chain	2.7–10
(f) hydrolysis of proteins and separation of the products	2.6.2
(g) the digestion of disaccharides and polysaccharides	3.2, 3.3, 3.5.2
(h) reducing and non-reducing sugars	3.4
(i) starch and cellulose in the human diet	3.5, 3.5.1–3
(j) the occurrence in plants and importance in the food industry of pectin,	10.2, 13.4
alginates,	13.4 and see **Additional Topic Food 1**
cellulose	3.5.3, 10.2
(k) the solubility and physical states of lipids	4.1
(l) iodine value of a lipid	4.7
(m) essential fatty acids	4.6
(n) the halogenation and hydrogenation of unsaturated fatty acids	4.7, 4.8
(o) vitamins; the consequences of inadequate intake of vitamins A and D	6.1, 6.2
(p) the importance of an adequate intake of minerals	7.2–6
(q) an investigation of nutrients present in food	see your practical work
(r) toxic metals in the food chain	2.15, 7.9
(s) the importance of water in the body	8.1, 8.2
(t) the colloidal and emulsion systems in milk and butter	8.3
(u) the calorific value of a food	1.3, 3.1, 4.9
(v) the energy and nutrient requirements of different types of people	1.3

2. Food additives

(a) E numbers	13.5
(b) the effects and uses of:	
tartrazine,	13.3, 13.5
caramel,	11.3.4
benzoic acid,	12.4, 12.4.7, 13.1
sulphur dioxide,	12.4, 12.4.7, 13.1
sodium nitrite,	11.4, 12.4.6
ascorbic acid,	6.2, 10.6.2, 11.2.2
butylated hydroxyanisole,	13.1
lecithin,	4.13, 4.14, 13.4
citric acid,	4.5, 10.6.2
monosodium glutamate	10.4, 13.2
(c) the addition of nutritional supplements	11.2.2, 14.3.7
(d) the ethics versus the economics of using additives in processed foods	**12.5** and **Chapter 13**
(e) flavours: complex mixtures of ingredients	10.4, 13.2
(f) identifying pigments in foods	10.6

3. Food quality

(a) quality of beef influenced by breed, gender, use of feed concentrates, growth promoters	9.8.2 and **Additional Topic Food 1**
(b)(c) structure of muscle tissue, texture, white and red meat	10.3, 11.4
(d) factors which affect the growth of fruit and vegetables	**Additional Topic Food 3**
(e)(f) crop production, selection of varieties	**Additional Topic Food 2**
(g) storing foods in a controlled environment; ripening of fruits	9.8.4 and **Additional Topic Food 4**

COMPLEMENTARY MODULE: FOOD TECHNOLOGY

4. Food storage		cheese,	12.3 under 'Bacteria', 12.4.9, 12.4.10, 12.3
(a) food contamination by micro-organisms	12.3		
(b) which may make food unpalatable, or reduce its nutritive value or cause food poisoning	12.3.1		
(c) microbial spoilage of foodstuffs, e.g. the souring of milk and the spoilage of strawberries by *Botrytis cinerea*	12.3 under 'Bacteria' and see **Additional Topic Food 5**	vinegar,	12.3 under 'Bacteria'
		baking	11.3
		(g) the enzymic browning of fruit; prevention	12.3.2
		(h)(i)(j) non-enzymic browning, e.g. in the potato crisp industry; prevention	10.6.2, 11.3.4 and see **Additional Topic Food 6**
(d) bacterial toxins compared with bacterial infestations, e.g. *Clostridium botulinum* and *Salmonella*	12.3.1, 12.4.3		
(e) the properties of enzymes	**Additional Topic How Far, How Fast? 5**	(k)(l) the formation of oxymyoglobin and metmyoglobin in fresh meat and methods of preventing metmyoglobin formation	11.4
(f) micro-organisms used in food production:		(m) the processes which cause rancidity in lipids	4.3, 4.5
5. Food processing and its effects on food		caramel	11.3.4
(a) the common methods of food processing	12.4	(e) starch and its reactions with water in food materials:	
(b) advantages of cooking foods:			
texture of meat	10.3, 11.4	starch,	3.5.1
and vegetables,	11.1	flour,	11.2
baking,	11.3	bread	11.3
destruction of micro-organisms	12.4.3, 12.4.9	(f) the effects of cooking on the texture and ascorbic acid content of potatoes	6.2, 6.2.1, 10.1, 10.6.2, 11.1
disadvantages of cooking foods:			
loss of vitamins in cooking water, and in bread-making	6.1, 6.2 6.3	(g) the effect of blanching on fruit and vegetable tissue	12.2, 12.4.1
loss of minerals in cooking water	7.1–3, 7.5, 7.6	(h) the terms:	
		anti-oxidant,	13.1
(c) denaturation of proteins	2.10	flavour enhancer,	13.2
(d) sugars in the food industry:		emulsifier,	13.4
flavouring,	10.4	preservative,	13.1
fruit preserving,	10.6.2	stabiliser and thickening agent	13.4
leavening,	11.3.2		

ADDITIONAL TOPIC FOOD TECHNOLOGY 1: QUALITY OF BEEF

Note that the references in this module are to *Biochemistry and Food Science* by E. N. Ramsden (Stanley Thornes, 1995).

The availability of high-quality beef depends on four factors: breeding, rearing, handling and butchery.

1. BREEDING

The dairy cow is the most important animal in the UK farming industry. Milk and dairy products represent 20% of the total farm output. The dairy cow is also a source of calves for the beef industry. After years of selective breeding the cow has a high milk yield and produces a large quantity of meat.

Until 1960, the predominant dairy breeds were Ayshires and Shorthorns, which did not produce particularly high-quality beef. Herefords and Aberdeen Angus were the most popular beef breeds. In 1961 the Government agreed to the importation of cattle from Europe, e.g. Charolais and Limousin, which give tender, tasty beef. Cross-breeding with these cattle has produced UK varieties which yield beef of high quality. A recent survey showed that consumers preferred meat from steers (castrated males) to heifers (cows that have not calved) or bulls. They preferred beef from suckler herds (cows which rear their own calves) to beef from dairy herds (cows used for milk production).

The quality of beef obtained from cows has been improved by cross-breeding.

2. REARING

Calves for beef production are born into a dairy herd and reared by a number of systems. Some are sold in the first few weeks to specialised beef producers. Others stay at the dairy farm until after weaning. All receive colostrum, the first milk from the mother, for the first few days. Then the calves are transferred on to a milk substitute fed from a bucket until they are weaned at five to six weeks. Horn buds are removed, and castration, when required, is performed. Bulls (uncastrated males) are reared in some systems of production, while steers (castrated males) are needed for others. After weaning the calves are fed concentrated foodstuffs, with hay or silage until three to six weeks old. There are three main finishing systems.

Cows for beef production are reared ...

(a) Cereal beef or barley beef

This is an intensive system which supplies a diet of **concentrates**. These are animal feedstuffs with a high content of protein and carbohydrates or fat, e.g. barley. They are usually supplied by manufacturers as balanced rations. The animals have this diet, and possibly a growth promoter, from three months until slaughter at 11–13 months. The cattle are housed all their lives so year-round production is possible. This method can be used to rear bulls – as well as steers and heifers – as 13-month-old bulls do not give the tainted meat of older bulls.

... as cereal beef, an intensive system which supplies a diet of concentrates ...

The animals are allowed to mature in ideal conditions, in which pests and competitors are eliminated. Only about 10% of the food given to animals is converted into flesh; the rest is used to keep the animals alive, maintain their temperature and enable them to move about. In intensive farming the energy loss is minimised by keeping the animals indoors at a constant temperature and restricting their freedom of movement. A larger fraction of the food they eat is converted into flesh.

(b) Silage beef

This is another intensive system of rearing which involves housing the animals. It uses silage as the major feed component. **Silage** is a conserved feed made from crops such as grass, kale, maize, which are cut green and preserved in a silo. The silage may be supplemented with concentrates. The animals do not grow as fast as barley-fed animals and are slaughtered at 16 months.

... as silage beef, an intensive system which supplies a diet of silage ...

(c) Eighteen-month beef

This is a semi-intensive system of rearing, in which the animals spend part of their time at pasture. Steers and heifers – not bulls – which have been born in the autumn are reared indoors for their first winter, then let out to pasture for six months and housed for a final six months for 'finishing' on a diet of silage and root vegetables.

... as eighteen-month beef, a system in which the animals spend a third of their lives at pasture.

Vaccination

Vaccination is especially important in intensive systems.

When large numbers of animals are reared close together, the spread of infectious micro-organisms is encouraged. Infectious diseases slow growth and reduce productivity. Treatment of an animal or a herd can be expensive, and the prevention of disease by vaccination is more effective.

Growth promoters

Growth promoters are used in some herds.

Vaccines keep animals healthy and promote growth. Healthy animals can be made still more productive if they are given hormones which stimulate growth and increase milk production. Growth hormones are proteins and can be produced by genetic engineering (**5.8**). The gene for the bovine growth hormone (bovine somatotrophin, BST) has been engineered into bacteria. As a result large quantities of the hormone are available for injection into cows. The results are startling. Milk production increases by 10–15%. The meat from treated cows carried less fat. There are, however, worries about the long-term effects of BST on human health. Analysis shows traces of BST in milk and meat, but BST is active only in cattle and is broken down in the human gut. It has been cleared for use in the USA but not in the UK.

3. HANDLING

If animals can be kept calm prior to slaughter the quality of the beef is better.

It is important to ensure minimal stress on the stock when they are selected for slaughter, loaded and transported to the abattoir. Arrival at the abattoir should be timed so as to minimise the time spent in holding pens before slaughter. This is especially important for bulls. Any stress or excitement, such as penning next to heifers or cows, will give the meat a higher pH and make it dark in colour and too firm and dry in texture.

4. BUTCHERY

See **10.3, 11.4**.

ADDITIONAL TOPIC FOOD TECHNOLOGY 2: CROP PRODUCTION

1. BREEDING

Over the thousands of years that human beings have been growing crops they have selected plants to meet their needs. Cereal grains are large and crop yields are higher than before humans intervened. Varieties have been produced which can resist the cold in winter and resist drought in summer, giving a longer growing season. Autumn-sown varieties of wheat and barley give higher yields than crops sown in the spring. Crops of today are chosen for high yields and ease of mechanical harvesting.

Cross-breeding has been used to improve crop yields ...
... the quality of the grain ...
... and resistance to disease

Plants are interbred to give new strains that combine the desirable characteristics of different varieties. A breeder might want to improve the resistance to disease of a commercial strain of wheat which gives high yields. Then pollen from a disease-resistant wild variety of wheat might be used to fertilise the commercial strain. The resulting plants would have genes from both parents and would have mixed characteristics, some good and some unwanted. The unwanted characteristics could be removed by repeated **back-crossing**: breeding with the high-yielding parent. Eventually a high-yielding disease-resistant plant would be produced.

Selecting, crossing and back-crossing can take many months and the breeding cycle can take many years. Now genetic engineering techniques (**5.8**) have been developed for

The development of new varieties is facilitated by genetic engineering techniques.

plants. Genetic engineers can take the desired genes from any source, not just from plants that can be interbred. The method is also much quicker than traditional breeding. The gene or genes isolated from the donor are inserted in a carrier molecule of DNA – the vector – which is inserted into the plant cell. There it gives rise to the protein coded for by the inserted gene.

2. GERMINATION

Seeds require favourable conditions in which to germinate. Most seeds contain very little water and their metabolism is slow. Before the metabolic changes of germination can begin, they must absorb water. The testa, the seed coat, is not very permeable to oxygen, and when it is soaked it allows oxygen to enter. Too much water may prevent germination by cutting down the oxygen supply to the seed.

Germination occurs most rapidly at temperatures up to about 40 °C. Above 45 °C the enzymes are denatured and the seedlings killed. Below 0–4 °C germination does not start at all in many seeds.

Seeds do not need light to germinate, but once the seedling is above ground it needs light to photosynthesise. The rate of photosynthesis is influenced by

- availability of water,
- temperature,
- mineral nutrition,
- competition from weeds,
- damage by insects.

Seeds must have favourable conditions to germinate.

See *Understanding Biology for Advanced Level* by G. Toole and S. Toole (Stanley Thornes Publishers) **23.4** or another A-level Biology text for an account of the factors affecting photosynthesis.

3. SOIL

Soil holds the water and mineral salts which plants need. The soil used for agriculture is improved mechanically by ploughing. Loosening the soil allows air to enter, allows water to drain and makes it easier for roots to penetrate. If water is scarce it can be added by irrigation. If the soil holds too much water it can be drained by digging ditches.

The soil must supply water and mineral salts.

Intensive methods of agriculture use large fields in which large machinery can be used effectively. Crops can be harvested more rapidly with less labour. Both irrigation and drainage are easier to manage in large fields.

4. FERTILISERS

Crop yields increased enormously as the use of nitrogenous fertilisers escalated. However, there comes a limit beyond which the use of excess fertiliser does no good. The use of nitrogenous fertilisers has remained constant during the last decade. Further increase in the yield of cereals will depend on plant breeding, rather than an increase in the use of fertilisers. The aim is to breed plants with the ability to channel nutrients into the grain, rather than into the straw, thus increasing the nutritional value of the crop.

Fertilisers are added to increase crop yields ….

5. HERBICIDES

… and herbicides reduce competition from weeds.

The introduction of selective herbicides in the late 1940s and 1950s began a breakthrough in raising crop yields. Weeds compete with the crop for moisture and nutrients. By controlling weeds, herbicides increase yields. In 1955 a farmer might harvest 2.5 tonnes of wheat per hectare; in 1995 the figure was 7 tonnes per hectare.

ADDITIONAL TOPIC FOOD TECHNOLOGY 3: FRUITS AND VEGETABLES

PRODUCTION IN THE FIELD

Crops of fruits and vegetables depend on the same factors as cereals.

The factors affecting crop production which have been discussed in Additional Topic Food Technology 2 apply to crops of fruit and vegetables as well as to cereals. **Plant growth substances**, e.g. gibberellins are used commercially to increase crops of fruits; see **4.10**.

GREENHOUSES

Greenhouses provide optimum conditions for plant growth ...

One way of providing optimum conditions is to grow crops under glass or under plastics, e.g. poly(ethene) film. In a greenhouse, the temperature can be regulated to the optimum temperature for the plant by heating in winter and shading in summer. The humidity can be controlled. Ventilators can be opened or closed to regulate the temperature. Ventilation also helps to replenish carbon dioxide levels. In greenhouses the technique of raising the level of carbon dioxide in the air has been used. At 1% carbon dioxide, growth is greatly accelerated. At carbon dioxide levels above this, the rate of photosynthesis is limited by the amount of sunlight. The technique is expensive and is only used when the produce can command a high price.

Sheets of plastic film are sometimes spread over plants growing in the field. This has been used, e.g. in Israel, for the production of winter vegetables. The higher temperature under the plastic encourages early development.

... and enable hydroponics to be used for some crops.

The technique of **hydroponics** is used in some greenhouses. Plants are grown in a nutrient solution which is circulated around the roots by pumps. No soil is needed; the roots are supported in a sand and gravel base. Hydroponics has been very successful in growing tomatoes, potatoes and rice with high yields and good flavours. It is useful in areas where the soil is poor, where water is limited and also to supply crops out of season.

ADDITIONAL TOPIC FOOD TECHNOLOGY 4: RIPENING FRUIT

Fruits and vegetables may be stored before sale.

Crops are stored to await the right time for sale. Conditions must be such as to prevent or reduce deterioration in quality. Losses are due to attack by insects and fungi. Cereals are stored in silos and grain stores. Potatoes and other root crops are stored in clamps. Fruits are stored in special fruit stores.

Ripening of fruits can be delayed by reducing the oxygen content of the air ...

Most fruits do ripen on the vine although some, e.g. bananas and pears, are normally picked green and ripened under controlled conditions. In many fruits the start of ripening is accompanied by an increase in respiration. Respiration provides energy for plants and this is needed as the plant matures. This increase is called the **climacteric rise**. Before the climacteric the plant is resistant to attack by diseases and microorganisms. It is partly for this reason that it is often advantageous to pick fruit green, inhibit ripening in some way while it is stored and then induce ripening under controlled conditions.

... or by increasing the level of carbon dioxide ... ... and by circulating air to remove the ethene produced by fruits when ripening starts.

Inhibition of ripening can be achieved by keeping the fruit in an atmosphere of 5–10% oxygen instead of normal air. In some cases a stream of air enriched with 10% of carbon dioxide is circulated to retard the climacteric rise in respiration. Refrigeration is an alternative method of slowing the climacteric phase.

In some plants the climacteric rise is accompanied by the emission of ethene. Ethene is known to speed the climacteric rise at concentrations as low as 1 ppm. A circulating air stream removes some of the ethene produced and delays the onset of ripening.

Ripening can be accelerated when the fruit is required for sale by releasing ethene into the air in the food store.

When the time comes when ripening is required, the concentration of ethene can be increased. At one time ethene was released from a cylinder into the store, but there was danger of gas being released too rapidly, with a detrimental effect on ripening fruit. Ethene is now produced at a controlled rate by passing ethanol vapour over a heated catalyst. It is a common practice to ripen bananas in an atmosphere enriched with ethene for 3–5 days. This form of ripening does not work well on green tomatoes: it makes them turn red without maturing.

ADDITIONAL TOPIC FOOD TECHNOLOGY 5: STRAWBERRIES SPOILED BY *BOTRYTIS CINEREA*

Protein-rich foods are attacked by bacteria ... ... carbohydrate-rich foods by fungi ... ... fruit and vegetables by fungi, bacteria and moulds, e.g. Botrytis cinerea *attacks strawberries.*

Fresh fruits and vegetables are rotted by fungi and bacteria. *Penicillium* causes blue mould rot of oranges. *Monilia* causes brown rot of apples, and *Botrytis cinerea* is the grey mould on strawberries. An infection of *Botrytis cinerea* causes water loss which increases the sugar content of the fruit. It is difficult for micro-organisms to enter undamaged fruit and vegetables, but when the surface is damaged spoilage organisms can quickly become established.

ADDITIONAL TOPIC FOOD TECHNOLOGY 6: POTATO CRISPS

Maillard browning gives foods an appetising smell.

Maillard browning (**11.3.4**) occurs at the surface of dry-cooked foods which contain protein and carbohydrate. Potato crisps are an example of such a food. The surface is brown with an appetising smell due to caramelisation of carbohydrates.

ANSWERS

ANSWERS TO CHECKPOINTS ON ADDITIONAL TOPICS

MODULE 4820: FOUNDATION

Checkpoint on Additional Topic 1
1. 5.25×10^{-3} mol KOH, 2.62×10^{-3} mol H_3PO_4
   $H_3PO_4 + 2OH^- \rightarrow HPO_4^{2-} + 2H_2O$
2. (a) 0.1 mol **X** : 0.2 mol Br_2 (b) 2 double bonds, $CH_2=CH-CH=CH_2$
3. $a=3, b=4, c=3, d=1, e=2$
4. (a) 0.25×10^{-3} mol $KMnO_4$, 1.25×10^{-3} mol Fe^{2+}
   (b) $MnO_4^-(aq) + 5Fe^{2+}(aq) + 8H^+(aq) \rightarrow$
   $Mn^{2+}(aq) + 5Fe^{3+}(aq) + 4H_2O(l)$
5. (a) Amount of $MnO_2 = 0.10$ mol, amount of $HCl = 0.40$ mol; therefore
   $MnO_2(s) + 4HCl(aq) \rightarrow MnCl_2(aq) + Cl_2(g) + 2H_2O(l)$
   (b) chlorine 2.40 dm^3 at rtp
6. (a) Amount of $Ce(IV) = 2.50 \times 10^{-3}$ mol, amount of $C_2O_4^{2-} = 1.25 \times 10^{-3}$ mol
   $C_2O_4^{2-}$ is oxidised to CO_2; therefore Ce(IV) is reduced to Ce(III)
   (b) $2Ce^{4+}(aq) + C_2O_4^{2-}(aq) \rightarrow 2Ce^{3+}(aq) + 2CO_2(g)$
   (c) Volume of $CO_2 = 2.5 \times 10^{-3} \times 24.0$ dm^3 at stp $= 60$ cm^3

Checkpoint on Additional Topic 2
1. In CH_4 the bonds are only slightly polar because carbon and hydrogen do not differ greatly in electronegativity (see **4.5.3**). Dipole–dipole interactions between molecules are not as strong as in more polar molecules (**4.7.1**). The energy needed to separate the molecules and change methane from the liquid state into the gaseous state is not great and T_b is low. In HCl the bonds are more polar than in CH_4 because chlorine and hydrogen differ more in electronegativity than carbon and hydrogen. The dipole–dipole interactions between molecules are strong, and more energy is needed to separate the molecules and T_b is therefore higher than T_b for CH_4. In HF, the bonds are highly polar and there is hydrogen-bonding in the liquid state (**4.7.3**). This makes it difficult to separate the molecules and raises T_b.
2. As the size of the halogen atom increases, and the number of electrons in the atom increases, van der Waals forces between the molecules increase (**4.7.2**) and the energy needed to melt the compound increases.
3. (a) The ions Mg^{2+} and O^{2-} are small ions with double negative charges, and the lattice energy of MgO is therefore higher than that of $Na^+ F^-$. MgO is therefore difficult to melt and has a high melting temperature.
   (b) From $F^- \rightarrow Cl^- \rightarrow Br^- \rightarrow I^-$ the size of the ion increases, and the $Na^+ X^-$ lattice is therefore less closely packed and the lattice enthalpy and the melting temperature decrease.
4. (a) dipole–dipole attraction (b) van der Waals forces (c) dipole–dipole attractions (d) hydrogen bonding

5. $SnCl_4$ is covalent with van der Waals forces between molecules, whereas $SnCl_2$ is an ionic compound.
6. Ar consists of single atoms. van der Waals forces are small.
7. CH_3OH Dipole–dipole interaction between C–O bonds and O–H bonds and hydrogen bonding between molecules.

Checkpoint on Additional Topic 3
1. (a) Ar, Cl_2, P_4, S_8
   (b) (i) Na, Mg (ii) none
   (iii) Al, Si, P, S, Cl, Ar
2. $MgCl_2$ consists of a three-dimensional ionic structure of Mg^{2+} ions and Cl^- ions with strong electrostatic forces of attraction between them. $SiCl_4$ consists of individual covalent molecules with only weak forces of attraction between molecules.
3.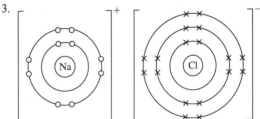

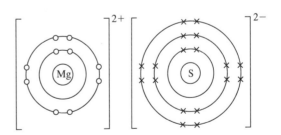

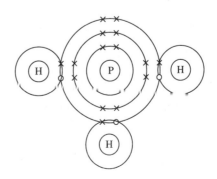

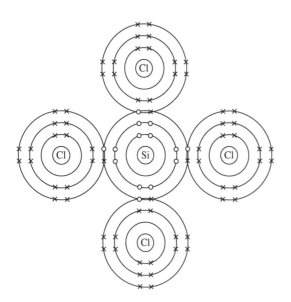

4. (a) Na, Mg (b) Si, P, S, Cl
   (c) Na_2O, MgO, Al_2O_3, SiO_2, P_2O_5, SO_3, Cl_2O_7; $NaOH$, $Mg(OH)_2$, $Al(OH)_3$, AlO_2^{2-}, $Si(OH)_4$, SiO_3^{2-}, H_3PO_4, H_2SO_4, $HClO_4$
5. See discussion in this topic.
6. See discussion in this topic.

MODULE 4821: CHAINS AND RINGS

Checkpoint on Additional Topic Chains and Rings 3
1. (a) Crude oil is fractionally distilled. Some fractions are cracked with the formation of ethene, which is polymerised.
   (b) Millions of years
   (c) No
   (d) Micro-organisms do not attack them. Plastic waste rots very slowly.
2. (a) Good thermal insulator
   (b) Possibly only minutes
   (c) Becomes non-biodegradable plastic waste.
   (d) Diffuses into the atmosphere. Eventually contributes to the greenhouse effect; this effect is less serious if the gas is a hydrocarbon than if it is a CFC.
3. (a) Items which are in use for a short time only, e.g. carrier bags, food cartons
   (b) Many examples of items which are in use for a long time, e.g. clothing, garden chairs
4. See the discussion in the topic.

MODULE 4822: TRENDS AND PATTERNS

Checkpoint on Additional Topic Trends and Patterns 2
1. Town bus services could use electric vehicles because they do not need to drive at high speeds or to cover long distances. They could run during the morning peak period, then return to the depot to have the batteries recharged and be ready to operate again during the evening peak period.
2. The mass of batteries is great compared with the mass of the car.
3. The reaction is not reversible. The efficiency is high: the cell voltage represents a high proportion of the enthalpy of the cell reaction. The energy density is much higher than other cells, but the power density is low (see Table 6.3).

Additional Topic Trends and Patterns 3
1. (a)

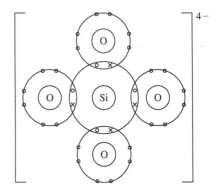

(b)

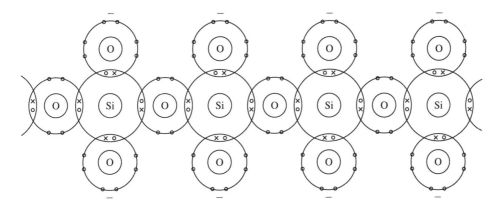

2. (a) Glasses fit the definition.
   (b) When molten silica is cooled rapidly it solidifies as a disordered glass instead of a crystalline solid.
   (c) See structures in Figures 2 and 3.
   (d) The amorphous structure of glass means that there are no regular planes of atoms to reflect light.
3. (a) Application of a force makes a metal change shape without fracturing; see **Figure 6.3** in 6.2.1. This cannot happen in a covalently-bonded ceramic which is a rigid three-dimensional structure.
   (b) Covalent bonds hold ceramics in a rigid structure, without the possibility of slip or movement of dislocations that metals have.
   (c) There are no free electrons as there are in the delocalised electron cloud in metals.

MODULE 4826: HOW FAR, HOW FAST?

Checkpoint on Additional Topic How Far, How Fast? 1
1. (a) The Li^+ ion in Period 2 is much smaller than the K^+ ion in Period 4; therefore the cation and anion are closer together in LiF and the force of attraction between them is greater.
   (b) The F^- ion in Period 2 is much smaller than the I^- ion in Period 5; therefore the cation and anion are closer together in NaF and the force of attraction between them is greater.
   (c) Cl^- has a single charge and O^{2-} has a double charge; therefore the force of attraction between Na^+ and O^{2-} is greater than the force of attraction between Na^+ and Cl^-.
2. MgO Both cation and anion have a double charge therefore the force of attraction between them is great.

Checkpoint on Additional Topic How Far, How Fast? 3
1. (a) 'Standard conditions' means the electrode is immersed in a 1 M solution of ions.
   (b) (v)
2. $MnO_4^-(aq) + 8H^+(aq) + 5e^- \rightleftharpoons Mn^{2+}(aq) + 4H_2O(l)$
   Hydrogen ions occur on the LHS of the equilibrium. Increasing $[H^+]$ therefore drives the equilibrium towards the RHS, in the direction of MnO_4^- acting as an oxidising agent.
3. $Cr_2O_7^{2-}(aq) + 14H^+(aq) + 6e^- \rightleftharpoons 2Cr^{3+}(aq) + 7H_2O(l)$
   Hydrogen ions occur on the LHS of the equibrium. Increasing $[H^+]$ therefore drives the equilibrium towards the RHS, in the direction of $Cr_2O_7^{2-}$ acting as an oxidising agent.

Additional Topic How Far, How Fast? 4
1. Rate = $k[NO_2Cl]$
2. (a) Rate = $k[A][B]^2$ (b) 3 (c) (i) ×2 (ii) ×4 (iii) ×8
3. It is possible that
   $2A \rightleftharpoons A_2$ Fast
   $A_2 + B \rightarrow C + D$ Slow
   For the equilibrium, k_f = rate constant of forward process, k_r = rate constant of reverse process
   Rate of forward reaction = rate of reverse reaction
   $\therefore k_f[A]^2 = k_r[A_2]$
   and $[A_2] = k_f[A]^2/k_r$
   and rate = $k_f[A]^2[B]/k_r$
   $= k[A]^2[B]$
4. (a) Rate = $k[NO][Br_2]$
   (b) Rate = $k'[NOBr_2][NO]$
   Since $NO + Br_2 \underset{k_r}{\overset{k_f}{\rightleftharpoons}} NOBr_2$
   $k_f[NO][Br_2] = k_r[NOBr_2]$
   and $[NOBr_2] = (k_f/k_r)[NO][Br_2] = k''[NO][Br_2]$
   Rate = $k'k''[NO][Br_2][NO]$
   $= k[NO]^2[Br_2]$
5. The step which involves CO is a fast step. The rate of reaction is controlled by the slow step.
6. (a) Order = 1 wrt $(CH_3)_3CBr$ because doubling the concentration doubles the rate.
   Order = 0 wrt OH^-; see experiments 1, 4 and 5
   Rate = $k[(CH_3)_3CBr]$
   (b) OH^- is not involved in the rate-determining step. Suggests that r.d.s. is
   $(CH_3)_3CBr \rightarrow (CH_3)_3C^+$ Br^-
   and OH^- reacts rapidly with $(CH_3)_3C^+$ once the cation has been formed to give the product.
7. Suggests that OH^- attacks C of CH_3Br. As the new HO—C bond forms, the old C—Br bond weakens: H—O···C···Br. The rate of reaction is proportional to the collision frequency.
8. Rate = $k_3[I^-][HOCl]$
   $k_1[OCl^-][H_2O] = k_2[HOCl][OH^-]$
   $[HOCl] = k_1[OCl^-]/k_2[OH^-]$
   Rate = $k_3k_1[I^-][OCl^-]/k_2[OH^-] = k[I^-][OCl^-]/[OH^-]$

Checkpoint on Additional Topic How Far, How Fast? 5
1, 2, 3, 4 See discussion in the topic.

Checkpoint on Additional Topic How Far, How Fast? 6
1. $RCO_2H + H_2O \rightleftharpoons RCO_2^- + H_3O^+$
   (a) Addition of hydrogen ions moves the equilibrium towards the LHS, to form more RCO_2H.
   (b) $OH^- + H_3O^+ \rightleftharpoons 2H_2O$
2. $RNH_2 + H_2O \rightleftharpoons RNH_3^+ + OH^-$
   (a) Added hydrogen ions combine with OH^- ions from

the base.
$OH^- + H_3O^+ \rightleftharpoons 2H_2O$
As OH^- ions are removed more are formed by the ionisation of RNH_2.
(b) Addition of OH^- drives the equilibrium to the LHS and more RNH_2 is formed.

3. (a) (i) $H_2NCH_2CO_2H + H_2O \rightleftharpoons H_2NCH_2CO_2^- + H_3O^+$
(ii) $H_2NCH_2CO_2H + H_2O \rightleftharpoons H_3N^+CH_2CO_2H + OH^-$
(b) $H_3O^+ + H_2NCH_2CO_2^- \rightleftharpoons H_2NCH_2CO_2H + H_2O$
(c) $OH^- + H_3N^+CH_2CO_2H \rightleftharpoons H_2NCH_2CO_2H + H_2O$
4. See discussion in this topic.

ANSWERS TO EXAMINATION QUESTIONS

Some questions have outline answers. Other questions have students' answers followed by comments from an examiner with corrections and suggestions for improvement.

QUESTIONS FROM MODULE 4820: FOUNDATION

Outline answers
1. (a) (i) See **25.7** (ii) See **25.8.3**
   (b) (i) See **25.8.2** (ii) See **26.3.2**
   (c) See **25.4.1**
   (d) (i) (ii) See **25.8.1**
2. (a) (i) $H_2S_2O_7 + H_2O \rightarrow 2H_2SO_4$
   (ii) $S \rightarrow SO_2 \rightarrow SO_3 \rightarrow H_2SO_4$
   98 g H_2SO_4 from 32 g S
   therefore 70 tonnes H_2SO_4 from
   $(70/98) \times 32$ tonnes = 23 tonnes S
   (iii) S: ox no = 0; SO_2 ox no of S = +4; SO_3 ox no of S = +6
   Stages $S \rightarrow SO_2$ and $SO_2 \rightarrow SO_3$ involve oxidation; step $SO_3 \rightarrow H_2SO_4$ does not.
   (b) amount of NaOH = 25.0×10^{-3} mol; amount of $H_2SO_4 = 12.5 \times 10^{-3}$ mol = $20.0 \times 10^{-3} \times M$
   $M = 6.25 \times 10^{-4}$ mol dm^{-3}
   Original concentration = $6.25 \times 10^{-4} \times 1000/50 = 12.5 \times 10^{-3}$ mol dm^{-3}
3. (a) He, Ne (b) Li, Na, K (c) He (d) Si
   (e) S (f) Si (g) P
4. (a) $^{23}_{11}$Na, 11p, 12n, 11e; $^{16}_{8}$O, 8p, 8n, 8e; $^{18}_{8}$O^{2-}, 8p, 10n, 10e
   (b) (i) Na, $1s^22s^22p^63s$ (ii) O^{2-}, $1s^22s^22p^6$
   (c) solid, high melting point, conductor of electricity when molten
   (d) $(0.75 \times 16) + (0.25 \times 18) = 16.5$
   (e) $44 = C^{16}O_2$, $46 = C^{16}O^{18}O$
5. (a) See **1.7**
   (b) Helium-3 has 2 protons + 1 neutron + 2 electrons in the atom. Helium-4 has 2 protons + 2 neutrons + 2 electrons in the atom.
   (c) (i) carbon-12
   (ii) $(0.992\% \times 3) + (99.008\% \times 4) = 3.990$
   (d) (i) see **2.2.2**
   (ii) He $\rightarrow$ He$^+$ + e$^-$
   (iii) The electron removed is in the 1s shell, the shell closest to the nucleus. This is true for hydrogen also, but the H nucleus has 2 protons to attract the electrons.
6. (a) See **4.2.1, 4.3, 4.6**
   (b) See *A-level Chemistry* **5.1** including **Figures 5.1, 5.6, 5.13, 5.12** (for I_3^- which has a similar arrangement of electrons pairs to SF_2)

Students' answers with Examiner's comments
7. (a) (i) 3 (ii) 5
   (b) (i) Amount of $NH_3 = 25.0 \times 10^{-3} \times 2.00$
   $= 50.0 \times 10^{-3}$ mol
   From the equation, amount of HNO_3 is the same, therefore

$V \times 0.500 = 50.0 \times 10^{-3}$ and
$V = 100 \times 10^{-3}$ dm$^3 = 100$ cm^3
(ii) Evaporate
(c) (i) M_r of $NH_4NO_3 = 28 + 4 + 48 = 80$
therefore sample = 0.1 mol
(ii) From the equation 0.1 mol N_2O is formed = 2.4 dm^3
(iii) $2N_2O \rightarrow 2N_2 + O_2$

Examiner's comments
7. (a) Carelessly, the student has omitted the sign in +3 and +5.
(c) The data are quoted to 3 significant figures. This candidate has lost marks by giving his answers as (i) 0.1 mol, instead of 0.100 mol and (ii) 2.4 dm^3 instead of 2.40 dm^3.
In (iii) s/he has omitted the state symbol (g).

---

8. (a) A hydrocarbon is a compound of hydrogen and carbon only. In a saturated hydrocarbon the carbon–carbon bonds are single bonds.
(b) Si_nH_{2n+2} and
(c) Structural isomers have the same molecular formulae but the arrangement of bonds is different.
(d)

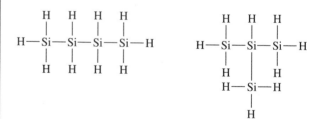

(e)

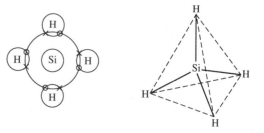

ANSWERS

> **Examiner's comments**
> 8. Some students were put off, thinking that they knew very little about silanes. This student has realised that s/he simply has to apply his or her knowledge of the structure of alkanes to the silanes to gain some easy marks.

9. (a) (i) $A \to B + C$. M_r of $C = 28$ therefore C has 2 carbon atoms and must be C_2H_4.
Subtracting C_2H_4 from the formula of A leaves C_8H_{18}, which must be the formula of B.
$B + Br_2 \to D + E$
The acidic gas could be HBr; then
$C_8H_{18} + Br_2 \to C_8H_{17}Br + HBr$
$C + Br_2 \to F$
$CH_2=CH_2 + Br_2 \to BrCH_2CH_2Br$
A is unbranched: $CH_3(CH_2)_8CH_3$, B is $CH_3(CH_2)_6CH_3$, C is $CH_2=CH_2$, D is $CH_3(CH_2)_6CH_2Br$, E is HBr, F is $BrCH_2CH_2Br$.

(ii) Unbranched-chain hydrocarbons burn with 'pinking' in vehicle engines. Cracking gives hydrocarbons with shorter chains. In reforming saturated hydrocarbons add to unsaturated hydrocarbons to form branched-chain hydrocarbons. These burn more smoothly with a higher octane number.

(b) In alkenes, RCH=CHR', addition takes place across the double bond. A second molecule of monomer can add across the double bond.

$$n \; \underset{H}{\overset{Cl}{>}}C=C\underset{H}{\overset{H}{<}} \longrightarrow \;\; \big(\!\!-\!\!\underset{H}{\overset{Cl}{\mid}}C-\underset{H}{\overset{H}{\mid}}C-\!\!\big)_n$$

Chloroethene   Poly(chloroethene)

$$n \; \underset{H}{\overset{H}{>}}C=C\underset{CO_2CH_3}{\overset{H}{<}} \longrightarrow \;\; \big(\!\!-\!\!\underset{H}{\overset{H}{\mid}}C-\underset{CO_2CH_3}{\overset{H}{\mid}}C-\!\!\big)_n$$

Methyl propenoate   Poly(methyl propenoate)

The polymers are not biodegradable. Buried in landfill sites, they do not rot and therefore need more and more land set aside for disposal.

(c) The reason why electrophiles do not attack benzene is that the π-electrons in the double bonds become delocalised – shared between all six carbon atoms. This gives the benzene ring stability – makes it less reactive.

> **Examiner's comments**
> 9. (a) (i) In a question like this, be careful to score enough points to gain all the marks on offer.
> (ii) Unlike many candidates, this student has not confused reforming with cracking.
> (b) S/he could have mentioned that this is a free radical addition reaction.
> (c) The student has not pointed out that as a result of delocalisation there is a lower electron density between two carbon atoms in the benzene ring than there is in C=C and the carbon–carbon bond in benzene, which is in between a single bond and a double bond in electron density, is less attractive to electrophiles.

10. (a) A fuel burns with the evolution of a large amount of energy per unit mass of fuel. The combustion products should not be toxic or pollutant.

(b) (i) The heat taken in, at constant pressure, when one mole of a compound is made from its elements under standard conditions.

(ii) $\Delta H = \Delta H_f$ of RHS $- \Delta H_f$ of LHS
$\Delta H = -277.7 - (-241.8) - (+52.3) = -88.2$ kJ mol^{-1}

(c) (i) A hydrogen atom bonded to an electronegative atom, e.g. O or N, forms a polar covalent bond: $-O^{\delta-}-H^{\delta+}$ or $-N^{\delta-}-H^{\delta+}$. The $\delta+$ H atom is attracted to a $\delta-$ atom in another molecule. This attraction is a hydrogen bond.
$-O^{\delta-}-H^{\delta+}\cdots\cdots O^{\delta-}-$

(ii)

$$H-\underset{H}{\overset{H}{\underset{|}{\overset{|}{C}}}}-\underset{H}{\overset{H}{\underset{|}{\overset{|}{C}}}}-O^{\delta-} \quad \overset{H^{\delta+}\cdots\cdots O^{\delta-}-H^{\delta+}}{H^{\delta+}}$$

> **Examiner's comments**
> 10. (a) The ignition temperature of the fuel must not be too high.
> (b) (i) It would have been better to add that 'standard conditions' means that gases are at a pressure of 1 atm and substances are in their normal states at the temperature stated.
> (ii) Many candidates lost marks here by neglecting $-$ signs in the values of ΔH.
> (c) (ii) Many candidates failed to show the dipole.
> (ii) Hydrogen bonding also takes place to the O atom of the alcohol:

$$H-\underset{H}{\overset{H}{\underset{|}{\overset{|}{C}}}}-\underset{H}{\overset{H}{\underset{|}{\overset{|}{C}}}}-O^{\delta-} \quad \overset{H^{\delta+}\cdots\cdots O^{\delta-}-H^{\delta+}}{H^{\delta+}} \\ \qquad\qquad\qquad \vdots \\ \qquad\qquad\qquad H^{\delta+}-O^{\delta-}\\ \qquad\qquad\qquad\qquad\;\; \backslash H^{\delta+}$$

11. (a) (i) mass spectrometry
    (ii) ^{123}Sb has 2 neutrons more in its nucleus than ^{121}Sb.
    (iii) $(57.25\% \times 121) + (42.75\% \times 123) = 121.86$
(b) (i) +3
    (ii) $2Sb_2S_3(s) \rightarrow 3CO_2(g)$
    10 mol $Sb_2S_3 \rightarrow 15$ mol $CO_2 = 15 \times 24$ dm^3 = 360 dm^3.

Examiner's comments
11. The structuring of this question makes it very easy to tell how much detail is required and makes it an easy question to score full marks on.

12. (a)

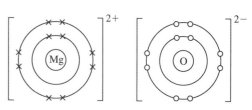

(b) (i) Na_2O (ii) SiO_2 (iii) MgO (iv) SO_3 (v) SO_3
(c) (i) $SO_3(s) + H_2O(l) \rightarrow H^+(aq) + HSO_4^-(aq)$
    (ii) $Na_2O(s) + H_2O(l) \rightarrow 2Na^+(aq) + 2OH^-(aq)$

QUESTIONS FROM MODULE 4821: CHAINS AND RINGS

Outline answers
1. (a) e.g. SOCl$_2$ (ii) NH$_3$
   (b) Heat under a reflux condenser.
   (c) propylamine
   (d)

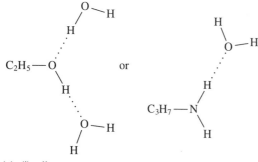

2. (a) (i) alkene group
       (ii) decolourises a solution of bromine without the evolution of HBr
   (b) (i) aldehyde group
       (ii) gives a reddish-brown precipitate with Fehling's solution (alkaline CuSO$_4$(aq))
   (c) (i) primary alcohol group
       (ii) oxidised by acid dichromate solution to a carboxylic acid
   (d) NaBH$_4$ in aqueous solution
3. (a) (i) **J** is 2, an aldehyde, oxidised by acidified dichromate and giving a 2,4-dinitrophenylhydrazone.
       (ii) **K** is 4, which gives a 2,4-dinitrophenylhydrazone but is not easily oxidised.
       (iii) **L** is 1: tertiary alcohols are difficult to oxidise.
       (iv) **M** is 3, a secondary alcohol which is oxidised to a ketone.
   (b) (i)

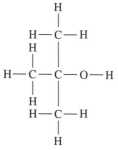

1,1-Dimethylethanol

Butanol

1-Methylpropanol

4. (a) (i) See **28.7.1, 28.7.2**
   (b) Chlorine in UV light adds to benzene to form 1,2,3,4,5,6-hexachlorocyclohexane. This is a free radical addition reaction.

Cyclohexene has only one double bond and adds one mole of Cl$_2$ per mole to form 1,2-dichlorohexane. This is electrophilic addition.

Benzene forms substitution products with chlorine and a Friedel Crafts catalyst, e.g. AlCl$_3$. This is electrophilic substitution.

Cyclohexene does not form substitution products with chlorine.

Students' answers with Examiner's comments

5. (a) 1 and 3
   (b) Reaction 2: chlorine in sunlight or UV light
   Reaction 3: chlorine + a halogen-carrier, e.g. $AlCl_3$
   (c) Chlorobenzene is unreactive because a lone pair of electrons on the Cl atom becomes part of the delocalised system of electrons in the benzene ring and this strengthens the Cl–ring bond.
   (d) In phenol, the lone pairs of electrons on the oxygen atom interact with the aromatic ring and increase the availability of electrons to attacking electrophiles.
   (ii) 2,4,6-tribromophenol

> **Examiner's comments**
> 5. (c) (d) The formulae would help here:
>
>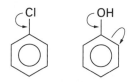

6. (a) **D** is C_6H_5CHO, **E** is $C_6H_5CO_2H$, **F** is $C_6H_5CH_2OH$, **G** is $C_6H_5CO_2Na$
   (b) (i) condensation of a carbonyl compound and 2,4-DNP to form a 2,4-dinitrophenylhydrazone
   (ii) oxidation of the aldehyde group –CHO to a carboxyl group –CO_2H
   (iii) reduction of an aldehyde group –CHO to an alcohol group, –CH_2OH
   (c) addition
   (d) (i)

   $$\begin{array}{c} OH \\ | \\ H-C-CN \\ | \\ C_6H_5 \end{array}$$

   (ii) optical

> **Examiner's comments**
> 6. (a) Note that with only four marks available, the student was right to state the identity of each compound and did not need to explain his or her reasoning.
> (c) 'Electrophilic addition' would have been better.
> (d) The student could have identified the chiral carbon atom in the formula.

7. (a) (i) flavourings, perfumes, solvents
   (ii) $CH_2OCOC_nH_{2n+1}$
   $|$
   $CHOCOC_nH_{2n+1}$
   $|$
   $CH_2OCOC_nH_{2n+1}$

The three groups C_nH_{2n+1} may be the same or different. Fats and oils are saponified (hydrolysed by alkalis) to give soaps. They are also an essential part of our diet.

(b) Bromoethane can be hydrolysed to ethanol.

$$C_2H_5Br(l) + H_2O(l) \rightarrow C_2H_5OH(aq) + HBr(aq)$$

Conditions: Heat bromoethane and concentrated aqueous sodium hydroxide under a reflux condenser for some hours. Separate ethanol by partition between water and ethoxyethane. Ethanol dissolves in the aqueous layer and bromoethane in the organic layer. Distil the aqueous layer to separate ethanol from water.
Ethanol can be oxidised to ethanoic acid.

$$C_2H_5OH(l) + 2[O] \rightarrow CH_3CO_2H + H_2O$$

Conditions: Warm ethanol with acidified potassium dichromate solution. Distil off ethanoic acid.
Ethanol and ethanoic acid form ethyl ethanoate.

$$C_2H_5OH(l) + CH_3CO_2H(l) \rightarrow$$
$$CH_3CO_2C_2H_5(l) + H_2O(l)$$

Conditions: Warm ethanol and ethanoic acid with a little concentrated sulphuric acid. Distil to obtain the low-boiling ester.

(c) Using an ester labelled with ^{18}O, if bond X is broken,

$$R-C\begin{array}{c}O\\\|\\\\^{18}O-R'\end{array} + H_2O \rightarrow$$

$$R-C\begin{array}{c}O\\\|\\\\O-H\end{array} + R'-^{18}O-H$$

If bond Y is broken,

$$R-C\begin{array}{c}O\\\|\\\\^{18}O-R'\end{array} + H_2O \rightarrow$$

$$R-C\begin{array}{c}O\\\|\\\\^{18}O-H\end{array} + R'-O-H$$

If ^{18}O is present in the alcohol, bond X has been broken. If ^{18}O is present in the acid, bond Y has been broken.

> **Examiner's comments**
> 7. This is a straightforward question but one that demands a good deal of knowledge, especially in (b). If you have simply learned your work in (a) and taken note of your practical work in (b) you will have scored well here.

8. (a) C: $64.9/12 = 5.41$, H: $13.5/1 = 13.5$, O: $21.6/16 = 1.35$
   The formula is $C_4H_{10}O$.
   Possibilities are three alcohols, C_4H_9OH and ethers, $C_2H_5OC_2H_5$ etc.
   K, which is oxidised to an acid, could be a primary alcohol, $CH_3CH_2CH_2CH_2OH$.
   J, which is also oxidised, could be a secondary alcohol being oxidised to a ketone; **J** could be $CH_3CH_2CHOHCH_3$.

H, which is not oxidised, could be a tertiary alcohol, $(CH_3)_3COH$ or it could be an ether, $C_2H_5OC_2H_5$.

(b) (i) **J** shows optical activity with a chiral carbon atom $CH_3CH_2C^*HOHCH_3$

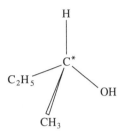

(ii) $CH_3CH_2CHOHCH_3 \rightarrow CH_3CH=CHCH_3 + H_2O$
$CH_3CH_2CHOHCH_3 \rightarrow CH_3CH_2CH=CH_2 + H_2O$
L and **M**: positional isomerism – the position of the double bond

(c) Phenol reacts with aqueous sodium hydroxide, whereas ethanol does not.
$C_6H_5OH(l) + NaOH(aq) \rightarrow C_6H_5ONa(aq) + H_2O(l)$

Examiner's comments
(a) Tertiary alcohols resist oxidation but are not unaffected by acidified potassium dichromate (VI).
(c) The student needs two more reactions. Esterification takes place less readily with phenols than with alcohols. The phenol must be converted into the phenoxide ion and then it will react with an acid chloride or anhydride.

$C_6H_5O^-Na^+ + (CH_3CO)_2O \rightarrow$
$\qquad C_6H_5OCOCH_3 + CH_3CO_2Na$

The replacement of $-OH$ by $-$Halogen takes place less readily in phenols than in alcohols. Phosphorus (V) chloride is used.

$C_6H_5OH(s) + PCl_5(s) \rightarrow$
$\qquad C_6H_5Cl(l) + POCl_3(s) + HCl(g)$

QUESTIONS FROM MODULE 4822: TRENDS AND PATTERNS

Outline answers
1. See Additional Topic Foundation 3. Remember that questions set on an optional module may include material from a compulsory module, such as the Foundation Module.

Students' answers with Examiner's comments
2. (a) Volatility decreases as the size of the atom increases. There are more electrons to give rise to temporary dipoles and to dipole–dipole interactions between molecules.
Thiosulphate is oxidised from $S_2O_3^{2-}$ to sulphate SO_4^{2-} by chlorine and bromine. Iodine is a weaker oxidising agent, and oxidises thiosulphate to tetrathionate $S_4O_6^{2-}$.
(b) The hydrides are covalent. When added to water, they act as acids, proton donors.

$HCl(aq) + H_2O(l) \rightarrow H_3O^+(aq) + OH^-(aq)$

The hydrides cannot act as proton donors unless a base, e.g. water, is present to accept a proton.
Acidic strength HCl < HBr < HI because bond strength HCl > HBr > HI
(c) $E^\ominus$ for $Cl_2 + 2e^- \rightarrow 2Cl^- = +1.36$ V
$E^\ominus$ for $Br_2 + 2e^- \rightarrow 2Br^- = +1.07$ V
$E^\ominus$ for $I_2 + 2e^- \rightarrow 2I^- = +0.54$ V
A reaction happens if $E^\ominus$ is positive, and since $E^\ominus_{Cl_2} > E^\ominus_{Br_2}$,

$Cl_2 + 2Br^- \rightarrow 2Cl^- + Br_2$.

Similarly, chlorine displaces iodine, and bromine displaces iodine.
(d) (i) Astatine is a weak oxidising agent because the strength of oxidising power decreases down the group. Astatine forms a strong acid HAt because the strength of the acids increases down the group. It forms salts such as NaAt and MgAt$_2$, which are crystalline solids which do not hydrolyse in water.
(ii) It is radioactive.

Examiner's comments
2. (a) Make the point that intermolecular forces are stronger as you go down the group.
It would be better to show the change in oxidation number. Thiosulphate is oxidised from $S_2O_3^{2-}$ (ox no of S = +2) to sulphate SO_4^{2-} (ox no of S = +6) by chlorine and bromine. Iodine is a weaker oxidising agent, and oxidises thiosulphate to tetrathionate $S_4O_6^{2-}$ (ox no of S = 2.5).
(c) State symbols should be included in the equations.

3. (a) $F = Le$
(b) (i) oxygen at the anode, hydrogen at the cathode
(ii) $2.00 \times 60 \times 60 \times 0.04 = 288$ C
(iii) $1.288/96\,500 = 2.98 \times 10^{-3}$ mol
(iv) $2H^+ + 2e^- \rightarrow H_2$
Amount of $H_2 = 2.98 \times 10^{-3}/2 = 1.49 \times 10^{-3}$ mol H_2
Volume $= 1.49 \times 10^{-3} \times 24.0$ dm$^3 = 35.8$ cm^3

Examiner's comments
3. This is a straightforward question with the mark scheme to guide you as to the amount of detail required. The student has been careful to write the equation for the cathode reaction and note that the amount of H_2 is half the amount of electrons.

4. (a) (i)

Element	Mg	Al	Si	P	S
Formula of oxide	MgO	Al$_2$O$_3$	SiO$_2$	P$_4$O$_{10}$	SO$_3$
Structure of oxide	Giant ionic	Giant ionic	Giant covalent	Molecular	Molecular

(ii) A: MgO dissolves sparingly to form an alkaline solution, Al_2O_3 and SiO_2 are insoluble; the oxides of P and S react to form strongly acidic solutions.
B: MgO, Al_2O_3 basic, react with acids to form salts of Mg^{2+} and Al^{3+}. Oxides of Si, P and S do not react.
C: MgO no reaction, Al_2O_3 reacts and dissolves to form a soluble aluminate, SiO_2 does not react with alkali, P_4O_{10} and SO_3 react to form a phosphate and a sulphate.

(b) Al^{3+} is smaller in size and more highly charged than Na^+ and Mg^{2+}, and Al_2O_3 therefore has the highest m.p.

Examiner's comments

4. (a) (i) The student would do better to describe P_4O_{10} and SO_3 as 'simple molecular' because giant covalent structures are also known as 'giant molecular'.
   (ii) These answers will score full marks; they are precise and contain much detail without being long. It is a good idea to look at the marks for each part: '3 marks' usually indicates three 'marking points'.
(b) This is partly answered. It misses the crucial point that both the small size and the large charge of Al^{3+} increase the forces of attraction within the ionic lattice.

5. (a) Fe $1s^2 2s^2 2p^6 3s^2 3p^6 4s^2 3d^6$ and
Cu $1s^2 2s^2 2p^6 3s^2 3p^6 4s^2 3d^8$
(b) (i) Fe^{2+} is formed when the two $4s^2$ electrons ionise. A 3d electron can ionise to form Fe^{3+}.
   (ii) With configuration Na $1s^2 2s^2 2p^6 3s$, only the 3s electron ionises. The energy of ionising a 2p electron is too high.
(c) (i) +1   (ii) +3

(d) (i) An ion and an oppositely charged ion or a neutral molecule or an atom and an ion combine by coordinate bonding to form a new substance.
   (ii) $[Cu(NH_3)_4]^{2+}$ $[CuCl_4]^{2-}$

Examiner's comments

5. (a) Be careful with questions like this. Many candidates do not read the question and answer both parts for atoms.
(b) It is not the electrons that ionise; it is the atoms.
(d) (i) This answer includes the important points of a coordinate bond from a ligand to a metal ion.
   (ii) $[Cu(NH_3)_4(H_2O)_2]^{2+}$ is preferred $[Cu(NH_3)_4]^{2+}$

6. (a) (i) yes   (ii) no
(b) $E^\ominus = E^\ominus_{Cu} - E^\ominus_{H_2} = +0.34 - 0 = +0.34$ V
   (ii) The energy of activation is too high.
(c) (i) $Cu^{2+} + 2e^- \rightarrow Cu$
   (ii) $Zn \rightarrow Zn^{2+} + 2e^-$
   (iii) $Zn + Cu^{2+} \rightarrow Cu + Zn^{2+}$
   (iv) From Zn to Cu
   (v) $E^\ominus_{cell} = E^\ominus_{Cu} - E^\ominus_{Zn} = +0.34 + 0.76 = 1.10$ V
   (vi) 1. E_{Cu} increases   2. E_{Zn} increases

Examiner's comments

6. (a) The student could have added that in (i) $E^\ominus$ is positive; in (ii) $E^\ominus$ is negative
(c) The state symbols should have been included in these equations.
(vi) 1. E_{Cu} increases therefore E_{cell} increases.
2. E_{Zn} increases – becomes less negative – therefore E_{cell} decreases.

QUESTIONS FROM MODULE 4826: HOW FAR, HOW FAST?

Outline answers

1. (a) (i) $K_c = [HI(g)]^2 / [H_2(g)][I_2(g)]$
   (ii) $55 = (3.67 \times 10^{-3})^2 / 5.10 \times 10^{-3} [I_2]$
   $[I_2] = 4.80 \times 10^{-5}$ mol dm^{-3}
(b) Le Chatelier's Principle. The system

$$H_2(g) + I_2(g) \rightleftharpoons 2HI(g); \quad \Delta H = +26.5 \text{ kJ mol}^{-1}$$

can absorb heat by moving from left to right; therefore K_c decreases with increasing temperature.
   (ii) No effect. K_c is constant at a constant temperature.
   (iii) No effect. The number of moles of gas is the same on LHS and RHS therefore the volume does not change and the equilibrium is not pressure-dependent.
2. (a) (i), (ii), (iii) see **14.1**.
(b) (i) see **14.1, 14.5, 14.11**
   (ii) A collision between 3 molecules is unlikely.

For a reaction $A + B \rightarrow X$
if the reaction step is first order in both **A** and **B**,
Rate $= k[A][B]$
If this is followed by $X + B \rightarrow C + D$
and this step is first order in **B** and in **X**,
Rate $= k'[B]$
Rate of formation of products $= kk'[A][B]^2$
and the reaction is third order.
3. (a) $C_2H_5OH(l) + 3O_2(g) \rightarrow 2CO_2(g) + 3H_2O(l)$
(b) (i) The enthalpy absorbed when one mole of a substance is completely oxidised under standard conditions.
   (ii) ethanol $-1370/46 = -29.8$ kJ mol^{-1}
   glucose $-3000/180 = -16.7$ kJ mol^{-1}
   (iii) ethanol $(30 \times 29.8) = 894$ kJ
   glucose $(20 \times 16.7) = 334$ kJ
   Total $= 1230$ kJ

(c) (i)

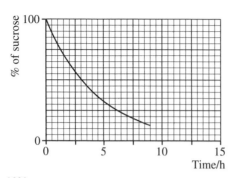

(ii) 10%

Students' answers with Examiner's comments

4. (a) (i) When $[S_2O_8^{2-}]$ doubles and $[I^-]$ is constant, the rate doubles: order wrt $S_2O_8^{2-}$ is 1. When $[I^-]$ doubles and $[S_2O_8^{2-}]$ is constant, the rate doubles, therefore order wrt I^- is 1. Rate $= k[S_2O_8^{2-}][I^-]$
$k =$ rate constant $= 4.10 \times 10^6/0.200 \times 0.0100 = 2.05 \times 10^{-3}$ dm^3 mol^{-1} s^{-1}

(ii) Iodine can be titrated against a standard solution of sodium thiosulphate.

$$I_2(aq) + 2S_2O_8^{2-}(aq) \rightarrow 2I^-(aq) + S_2O_6^{2-}(aq)$$

The rate constant increases with temperature.

(iii) Draw a tangent to the curve at zero time.

(b) Rate $= k[^{14}C]$
Number of half-lives $= 3$, therefore age $= 16\,800$ years

(c) (i) The variable oxidation state enables them to give or accept electrons.

(ii) A heterogeneous catalyst is in a different phase from the reactants, e.g. nickel catalysing the reaction between hydrogen and the vapour of an unsaturated hydrocarbon.
A homogeneous catalyst is in the same phase as the reactants, e.g. the enzyme catalase catalysing the decomposition of hydrogen peroxide.

Examiner's comments

4. (a) In comparing 8.20×10^{-6} with 1.64×10^{-5} it would have been safer to convert both to the same power: 16.4×10^{-6}.
The student has been careful to work out rate constant units correctly. You cannot memorise rate constant units; they must always be worked out from first principles.
S/he has omitted units from the graph; the concentration unit is mol dm^{-3} and the time unit is s.

(c) This part, testing standard content, was well answered so make sure that you too learn the work!

5. (a) (i) A Brönsted–Lowry acid is a substance which can give a proton to another substance.
Hydrogen chloride and ethanoic acid can both donate protons to water.

$$HCl(aq) + H_2O(l) \rightarrow H_3O^+(aq) + Cl^-(aq)$$

$$CH_3CO_2H(aq) + H_2O(l) \rightleftharpoons CH_3CO_2^-(aq) + H_3O^+(aq)$$

In the case of hydrochloric acid, ionisation is complete. This is why hydrochloric acid is a strong acid. In the case of ethanoic acid only a fraction of ethanoic molecules dissociate, and the concentration of hydrogen ions in solution is low. This is why ethanoic acid is a weak acid.

(ii)

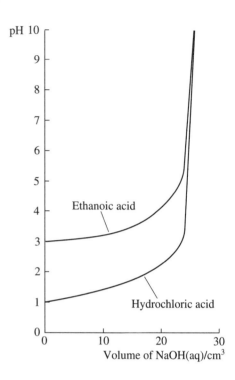

(iii) Methyl orange changes colour over a pH range when the pH is changing rapidly just before and just after the end point in the hydrochloric acid titration. It can be used. In the ethanoic acid titration, methyl orange changes colour before the rapid change in pH at the end point begins so it cannot be used.

(b) Amount of
NaOH $= 10.0 \times 10^{-3} \times 2.0 = 20 \times 10^{-3}$ mol
$2NaOH(aq) + H_2SO_4(aq) \rightarrow Na_2SO_4(aq) + H_2O(l)$
Amount of $H_2SO_4 = 10 \times 10^{-3}$ mol
$=$ Volume $\times 0.20$ mol dm^{-3}
Volume $= 10 \times 10^{-3}/0.20 = 50 \times 10^{-3}$ dm$^3 = 50$ cm^3

(c) $K_{In} = [H^+][In^-]/[HIn] = 1.0 \times 10^{-4}$
$[H^+]^2 = 1.0 \times 10^{-4} \times 0.10$
$[H^+] = 3.16 \times 10^{-3}$ and pH $= 2.5$

Examiner's comments

5. (a) (ii) Make sure that the titration curve starts at the correct pH, in this case pH ≈ 3 for a weak acid, and finishes at the correct pH, in this case pH ≈ 14 for a strong alkali. Make sure the pH changes rapidly after the correct volume of alkali.

(iii) You should add here that the titration curve for ethanoic acid shows that an indicator is needed that changes colour at pH ≈ 6–12.

(b) In calculations, show your working. Correct working can be given credit, but an incorrect answer on its own will score nothing at all.

ANSWERS

6. (a) The overall properties of the system do not change, but changes are taking place all the time, from left to right of the equilibrium and from right to left.
   (b) (i) When conditions change, the system adjusts to minimise the effects of the change. In this reaction, three volumes of gas form two volume of gas. According to Le Chatelier's Principle increasing the pressure will make the gases contract, and they can do this if more SO_2 and O_2 combine to form SO_3. The reaction is exothermic so raising the temperature will make the equilibrium move from right to left to absorb heat.
   (ii) High pressure is not used because SO_2 condenses at high pressure. A low temperature is not used because the system takes a long time to reach equilibrium at low temperature. A catalyst speeds up the reaction.
   (iii) $$K_p = p_{SO_3}^2/(p_{SO_2}^2 \times p_{O_2})$$
   $$= (4.5)^2/(0.90)^2 \times 0.083 = 3.0 \times 10^4 \text{ atm}^{-1}$$
   (c) (i) catalyst
   (ii) $SO_3 + H_2O \rightarrow H_2SO_4$
   Buildings are attacked through acid reacting with limestone, marble, concrete and metals.
   Crops are attacked when soil becomes acidic.
   (d) (i) $2NO \rightarrow N_2 + O_2$
   $2NO_2 \rightarrow N_2 + 2O_2$
   (ii) It increases the area of the surface on which oxides can react.
   (iii) Lead compounds poison the catalyst.

Examiner's comments
6. (b) (i) This answer is nearly there. A top response would state that the increased pressure is relieved by the reaction reducing the number of moles of gas present (3 volumes → 2 volumes).
   (ii) The key points required here are the *cost* of generating high pressures and the *dangers* that high pressures bring to pipes etc. You should always discuss the compromise required for optimum conditions, i.e. a high enough temperature for a reasonable rate but not too high to move the equilibrium completely to the left.
   (iii) The student has got the unit of K_p correct. You can't memorise the units of K_p; you must work each case out from first principles.

7. (a) (i) $N_2(g) + 3H_2(g) \rightleftharpoons 2NH_3(g)$
   (ii) It is an equilibrium reaction. An increase in pressure makes the system adjust by decreasing in volume, and it does this by going from left to right.
   (iii) The position of equilibrium is attained faster. The reaction is exothermic so the equilibrium can adjust to an increase in temperature by going from right to left.
   (b) (i) $$K_p = \frac{(p_{NH_3})^2}{(p_{N_2})(p_{H_2})^3}$$
   (ii) $K_p = (75)^2/35 \times (90)^3 = 2.20 \times 10^{-4} \text{ atm}^{-2}$
   (iii) atm^{-2}
   (iv) No effect on K_p
   (c) An enzyme has an active site. This is a dent or cleft in the molecule that the substrate can fit into. It contains groups which bond to the substrate. Only one substrate or a few similar substrates are the right size and shape to fit into the active site and have the right groups to bond to it.
   (d) $\Delta H = \Delta H(CO_2) + 2\Delta H(NH_3) - \Delta H(CO(NH_2)_2)$
   $\quad - \Delta H(H_2O)$
   $= -414.5 + 2(-81.0) - (-320.5) - (-287.0)$
   $= 31.0 \text{ kJ mol}^{-1}$

Examiner's comments
7. (a) Le Chatelier's Principle.
   (b) You cannot memorise the unit of K_p; this must be worked out from first principles in each case.

8. (a) (i) Compare experiments 2 and 3, in which $[C_2H_5OH]$ is the same. When $[(CH_3CO)_2O]$ doubles the rate doubles; therefore the reaction is first order in ethanoic anhydride.
   (ii) Compare experiments 1 and 2. The rate of experiment 2 is four times that of experiment 1. When $[(CH_3CO)_2O]$ doubles the rate doubles; therefore the rate must be doubled again by doubling $[C_2H_5OH]$. The reaction is first order in ethanol.
   (b) (i) Rate $= k[(CH_3CO)_2O] [C_2H_5OH]$
   (ii) 2
   (iii) $3.30 \times 10^{-4} \text{ mol dm}^{-3} \text{ s}^{-1}$
   $= k \times 0.200 \text{ mol dm}^{-3} \times 0.200 \text{ mol dm}^{-3}$
   $k = 8.25 \times 10^{-3} \text{ mol dm}^{-3} \text{ s}^{-1}$
   (iv) Ethanol is in excess so $[C_2H_5OH]$ remains approximately constant throughout the reaction.
   Rate $= k[(CH_3CO)_2O)] \times$ constant
   $= k'[(CH_3CO)_2O)]$

Examiner's comments
8. This is an easy question on which to score full marks. Work through each step carefully, showing your working in the calculation. In (b) (iii) it is a good idea to include the units in your calculation so that you obtain the correct unit for k.

Index of Additional Topics

Avogadro constant, determination 40

Batteries, improved 41
Beef quality 103–4
Bonding 11
Botrytis cinerea 108
Buffer solutions, biological importance 83–4

Ceramics 43
Crop production 105–7
Cytochrome 95

Electrochemical transport 42
Electrode potential, effect of concentration 77

Enzymes 80–2
Esters, commercial uses 31

Fruit ripening 107–8
Fruits 107

Glasses 45
Glucokinase 95

Haemoglobin 95
Hexokinase 95

Lattice energy, effect of ionic charge and radius 75

Period 3 elements, compounds 15–16
   properties 11–12
   reactions 13–14
Potato crisps 108
Proteins, hydrolysis 28

Reaction mechanisms 78
Recycling
   glass 31
   metals 29
   plastics 29

Standard electrode potential, measurement 76
Stoichiometry 8

Vegetables 107